AF880119

Continuous Extracorporeal Treatment in Multiple Organ Dysfunction Syndrome

Contributions to Nephrology

Vol. 116

Series Editors
G.M. Berlyne, Brooklyn, N.Y.
S. Giovannetti, Pisa

Editorial Board
V. Bonomini, Bologna
J. Churg, New York, N.Y.
K.D.G. Edwards, New York, N.Y.
E.A. Friedman, Brooklyn, N.Y.
S.G. Massry, Los Angeles, Calif.
R.E. Rieselbach, Milwaukee, Wisc.
J. Traeger, Lyon

KARGER

Basel · Freiburg · Paris · London · New York ·
New Delhi · Bangkok · Singapore · Tokyo · Sydney

3rd International Conference on Continuous Hemofiltration,
Vienna, July 8, 1994

Continuous Extracorporeal Treatment in Multiple Organ Dysfunction Syndrome

Volume Editors *H.G. Sieberth,* Aachen
H.K. Stummvoll, Linz
H. Kierdorf, Aachen

39 figures and 39 tables, 1995

KARGER

Basel · Freiburg · Paris · London · New York ·
New Delhi · Bangkok · Singapore · Tokyo · Sydney

....

Contributions to Nephrology

Library of Congress Cataloging-in-Publication Data
International Conference on Continuous Hemofiltration (3rd: 1994: Vienna, Austria)
Continuous extracorporeal treatment in multiple organ dysfunction syndrome / volume editors,
H.G. Sieberth, H.K. Stummvoll, H. Kierdorf.
(Contributions to nephrology; v. 116)
'3rd International Conference on Continuous Hemofiltration, Vienna, July 8, 1994.'
Includes bibliographical references and index. (hard cover)
1. Multiple organ failure – Congresses. 2. Continuous arteriovenous hemofiltration – Congresses.
3. Hemodialysis – Congresses.
I. Sieberth, Heinz-Günter. II. Stummvoll, H.K. III. Kierdorf, H., 1957– . IV. Title. V. Series.
RB150.M84I58 1995
616′.047 – dc20
ISBN 3-8055-6131-8

Bibliographic Indices. This publication is listed in bibliographic services, including Current Contents® and Index Medicus

Drug Dosage. The authors and the publisher have exerted every effort to ensure that drug selection and dosage set forth in this text are in accord with current recommendations and practice at the time of publication. However, in view of ongoing research, changes in government regulations, and the constant flow of information relating to drug therapy and drug reactions, the reader is urged to check the package insert for each drug for any change in indications and dosage and for added warnings and precautions. This is particularly important when the recommended agent is a new and/or infrequently employed drug.

All rights reserved. No part of this publication may be translated into other languages, reproduced or utilized in any form or by any means, electronic or mechanical, including photocopying, recording, microcopying, or by any information storage and retrieval system, without permission in writing from the publisher.

© Copyright 1995 by S. Karger AG, P.O. Box, CH–4009 Basel (Switzerland)
Printed in Switzerland on acid-free paper by Thür AG Offsetdruck, Pratteln
ISBN 3-8055-6131-8

Contents

Preface IX

Acute Renal Failure in Multiple Organ Dysfunction Syndrome

Pathobiology of the Endothelium in Multiple Organ Dysfunction Syndrome 1
Kirkpatrick, C.J.; Klein, C.L.; Bittinger, F. (Mainz)

Mediators in Continuous Extracorporeal Treatment of Multiple Organ Dysfunction Syndrome 10
Sunder-Plassmann, G.; Hörl, W.H. (Vienna)

Prognostic Importance of Vital Organ Dysfunction in Sepsis 16
Suter, P.M.; Pittet, D. (Geneva)

Intensive Care Management of Patients with Acute Hepatic and Renal Failure . . . 22
Davenport, A.; Kirby, S.A. (Bristol)

Different Forms of Continuous Extracorporeal Treatment: Technical Aspects

Continuous Renal Replacement Therapies: The Need for a Standard Nomenclature . 28
Ronco, C. (Vicenza); Bellomo, R. (Heidelberg, Vic.)

Comparison between Continuous Venovenous and Intermittent Hemodialysis in Acute Renal Failure 34
Friedrichsohn, C.; Pollak, A.; Fischer, F.; Köhler, H.; Riegel, W. (Homburg/Saar)

Continuous Venovenous Hemofiltration in Acute Renal Failure: Is a Bicarbonate- or Lactate-Buffered Substitution Better? 38
Kierdorf, H.; Leue, C.; Heintz, B.; Riehl, J.; Melzer, H.; Sieberth, H.G. (Aachen)

Removal of Methotrexate by Continuous Venovenous Hemodiafiltration 48
Jambou, P.; Levraut, J.; Favier, C.; Ichai, C.; Milano, G.; Grimaud, D. (Nice)

Hemofiltration in a Nephrotic Syndrome of Unknown Origin 53
Reiger, J. (Stolzalpe)

MODS, Sepsis and Their Mediators during Continuous Extracorporeal Treatment

Influence of Venovenous Hemofiltration on Posttraumatic Inflammation and Hemodynamics 56
Riegel, W.; Ziegenfuss, T.; Rose, S.; Bauer, M.; Marzi, I. (Homburg/Saar)

Cytokine Kinetics (TNF-α, IL-1β, IL-6) during Continuous Hemofiltration: A Laboratory and Clinical Study 62
Bommel, E.F.H. van; Hesse, C.J.; Jutte, N.H.P.M.; Zietse, R.; Bruining, H.A.; Weimar, W. (Rotterdam)

Hemofiltration in Human Sepsis: Evidence for Elimination of Immunomodulatory Substances 76
Hoffmann, J.N.; Faist, E.; Deppisch, R.; Hartl, W.H.; Inthorn, D. (Munich)

Complement Fragments and Cytokines: Production and Removal as Consequences of Hemofiltration 80
Journois, D. (Paris)

Ultrafiltration Allows to Reduce Cytokine Plasma Concentrations during Pediatric Cardiopulmonary Bypass 86
Journois, D.; Pouard, P.; Rolland, B.; Lagarde, M. (Paris)

Effect of Continuous Hemodiafiltration on IL-6, TNF-α, C3a, and TCC in Patients with SIRS/Septic Shock Using Two Different Membranes 89
Braun, N.; Rosenfeld, S. (Tübingen); Giolai, M. (St. Gallen); Banzhaf, W.; Fretschner, R.; Warth, H.; Weinstock, C.; Deppisch, R.; Erley, C.M.; Müller, G.A.; Risler, T. (Tübingen)

The Influence of Continuous Hemofiltration on Cytokine Elimination and the Cardiovascular Stability in the Early Phase of Sepsis 99
Sander, A.; Armbruster, W.; Sander, B.; Philipp, Th.; Schäfer, Ch.; Thürauf, N.; Lange, R. (Essen)

Continuous Hemofiltration in Phase of Recovery from Septic Shock 104
Redl-Wenzl, E.M.; Sporn, P. (Vienna)

Effects of Continuous Venovenous Hemofiltration on Pulmonary Function and Hemodynamics in Postoperative Septic Multiorgan Failure 108
Krüger, I.; Jacobi, C.; Landwehr, P. (Cologne)

Continuous Monitoring of Cardiac Output during Hemodialysis 112
Bastien, O.; Filley, S.; Paulus, S.; Piriou, V.; Estanove, S. (Lyon)

Continuous Treatment in Patients with Normal Renal Function

Continuous Extracorporeal Treatment in Patients with Normal Renal Function? . . . 115
Blinzler, L.; Bödeker, H.; Gebhardt, C.; Heuser, D. (Nuremberg)

Nonrenal Indications for Continuous Hemofiltration Therapy in Patients with Normal Renal Function? 121
Druml, W. (Vienna)

Anticoagulation

Low-Dose Heparinization for Anticoagulation in Intensive Care Patients on Continuous Hemofiltration 130
Gretz, N.; Quintel, M.; Ragaller, M.; Odenwälder, W.; Bender, H.J.; Rohmeiss, P.; Strauch, M. (Mannheim)

Use of Prostacyclin in Patients with Continuous Hemofiltration after Open Heart Surgery 136
Klotz, K.-F.; Gellersen, H.; Brzelinski, T.; Sedemund-Adib, B.; Gehring, H. (Lübeck)

Influencing Hemostasis during Continuous Venovenous Hemofiltration after Acute Renal Failure: Comparison with Intermittent Hemodialysis 140
Stefanidis, I.; Hägel, J.; Kierdorf, H.; Maurin, N. (Aachen)

Influence of Coagulation Parameters on Filter Running Time during Continuous Venovenous Hemofiltration 145
Stefanidis, I.; Hägel, J.; Maurin, N. (Aachen)

The Mechanism of Postoperative Blood Loss Reduction by Ultrafiltration in Cardiac Surgery Is Unlikely to Be Only Mediated by Hemoconcentration of Coagulation Factors 150
Journois, D.; Raucourt, E. de.; Fischer, A.M.; Allanic, G.; Keita, M.; Courcy, N. de (Paris)

Antithrombin III Deficiency during Continuous Venovenous Hemodialysis 154
Bastien, O.; French, P.; Paulus, S.; Filley, S.; Berruyer, M.; Dechavanne, M.; Estanove, S. (Lyon)

Thromboelastography and Monitoring of Coagulation in Patients Undergoing Continuous Venovenous Hemofiltration 159
Derrier, M.; Jambou, P.; Kaidomar, M.; Ichai, C.; Bayle, J.; Grimaud, D. (Nice)

Pediatric Application

Prognosis in Pediatric Patients with Multiple Organ System Failure and Continuous Extracorporeal Renal Support 163
Zobel, G.; Ring, E.; Rödl, S. (Graz)

Early Experience with a New Hemodiafilter for Small Infants 169
Zobel, G.; Ring, E.; Rödl, S. (Graz)

Continuous Venovenous Hemofiltration in a Seven-Year-Old Child Suffering from Severe Sepsis and Multiple Organ Failure 173
Lampert, R.; Weih, E.H.; Hikl, R.; Mazuch, M. (Wuppertal)

Removal of Complement Fragments by Ultrafiltration during Pediatric Cardiopulmonary Bypass 179
Journois, D.; Vaccaroni, L.; Rambur, R.; Lagarde, M.; Keita, M.; Pouard, P. (Paris)

Subject Index 183

Preface

The rate of mortality due to acute renal failure has not changed in the last 20 years, although further developments in intensive care and extracorporeal treatment have been made. This is due to the fact that acute renal failure in the nineties in nearly all cases is no single organ failure but part of a multiple organ dysfunction syndrome. This syndrome modifies the typical clinical presentation of acute renal failure.

Since 1977, when Kramer et al. published the first report of continuous extracorporeal treatment, and since the first symposium on continuous hemofiltration, held in Aachen in 1984, the main topics in the field of continuous extracorporeal treatment had changed.

A number of continuous treatment modalities (i.e. pump-driven venovenous methods, continuous dialysis methods) have been developed, allowing sufficient control of azotemia also in hemodynamically unstable patients with hypercatabolism. A lot of mediators and cells have been identified to be involved in the development of multiple organ dysfunction syndrome. Especially plasma cascades, mediators derived from granulocytes and endothelial cells play a major role and are of great scientific interest in the nineties.

One of the main topics of the Third International Conference on Continuous Hemofiltration was therefore the standpoint of continuous techniques in the treatment, development, origin and outcome of multiple organ dysfunction syndrome. Many papers were presented, dealing with elimination, adsorption or induction of inflammatory mediators (i.e. TNF, interleukin-1) involved in the disease. As continuous anticoagulation is still needed in the most continuous treatment modalities, 'alternative' forms of anticoagulation were a second topic of the conference. The last 'highlight' of the symposium was a half-day controversial discussion about nonrenal indications for continuous forms of treatment.

As the Second Conference in 1991, the symposium was a part of the yearly conference of the European Dialysis and Transplantation Association. This is due to the fact that the differentiated transport properties of the more complicated continuous treatment modalities, i.e. diffusive and convective transport combined with adsorptive processes and the technical requirements, remain the domain of nephrologists with specific experience in intensive care, or intensive-care specialists with specific experience in nephrology. Continuous treatment modalities are more than methods to eliminate water and urea.

H.G. Sieberth
H.K. Stummvoll
H. Kierdorf

Sieberth HG, Stummvoll HK, Kierdorf H (eds): Continuous Extracorporeal Treatment in Multiple Organ Dysfunction Syndrome. Contrib Nephrol. Basel, Karger, 1995, vol 116, pp 1–9

Pathobiology of the Endothelium in Multiple Organ Dysfunction Syndrome

C. James Kirkpatrick, Christoph L. Klein, Fernando Bittinger

Institute of Pathology, University Clinic of the Johannes Gutenberg University, Mainz, Germany

Multiple organ dysfunction syndrome (MODS) describes the broad spectrum of symptoms and signs which form the background on which multiple organ failure (MOF) develops. In many cases, MODS develops after a certain latency period in the context of a host reaction to a primary insult (e.g. trauma, burns injury, pancreatitis), this reaction being generally termed SIRS (systemic inflammatory response syndrome) [1]. A further significant component is the frequent complication of sepsis or even septic shock. Although these processes involve the activation of numerous mediator cascades, the present paper will concentrate on the role of the endothelium, especially in the microcirculation, in responding to these mediators. Space constraints preclude a detailed discussion, which is presented elsewhere [2].

Microcirculatory Dysfunction: Morphological Evidence

Morbid anatomical studies (at light and electron microscopical level) of the adult respiratory distress syndrome (ARDS), which forms an element of MODS, gave the first real insight into the central role of the microcirculation [3]. Hydropic swelling, followed by necrosis of the endothelial cells (EC) can be found in the capillary network. Impairment of perfusion is further complicated by neutrophilic granulocyte sequestration and microthrombus formation in the lumen of the microcirculation. A cursory glance at these lesions might suggest that the humoral and cellular compartments of the blood would suffice to explain these pathological changes. This impression is misleading. The major advancement in our understanding of the pathognesis of MODS during the past 10 years is the recognition of the essential role of the endothelium in maintaining and amplifying the microcirculatory disturbances.

Microcirculatory Dysfunction: Pathophysiological and Cell Biological Aspects

The concept of the endothelium as a passive barrier has been replaced by the concept of a multifunctional interface with numerous sensory and modulator functions, which make it a key player in various physiological and pathological responses. From a pathophysiological viewpoint, the essential sequence in MODS appears to be peripheral vascular insufficiency, leading to maldistribution of flow, which is combined with defective oxygen extraction. The cell pathobiological hypothesis concerning microcirculatory dysfunction is based on the multifunctional nature of the endothelium and regards events centering on this entity *prior* to the morphological changes (described above) as of major pathogenetic significance. In this view the quintessence is, therefore, that *endothelial dysfunction* is a principal pathogenetic event in the syndrome complex involving SIRS, sepsis and MODS.

To understand this concept, it is necessary to analyze the broad spectrum of functions with which the endothelium is equipped, a state of major teleological importance, as this flexibility enables the endothelium to react appropriately to injury. However, in the context of MODS, this same adaptability can spell disaster [4]. It is relevant to regard the endothelium as a Janus face, with the following opposing elements: (1) anti- and prothrombogenic activity; (2) vasodilating and vasoconstricting functions; (3) modulation and amplification of inflammation, and (4) growth-inhibitory and growth-promoting functions.

Anti- and Prothrombogenic Activity

Table 1 lists the most important anti- and prothrombogenic substances produced by the endothelium. Space restrictions do not permit a detailed presentation, although some important aspects should be emphasized. The physiological situation in situ involves a predominance of the anticoagulatory activity of the endothelium [5]. It has been known for a long time that EC can rapidly synthesize and secrete prostacyclin [6], a potent inhibitor of platelet aggregation, a function also shared by NO [7]. In MODS, and in particular those forms associated with sepsis, this delicate hemostatic regulatory activity is tipped in favor of the proco-

Table 1. Anti- and procoagulant activity of EC

Anticoagulant	Procoagulant
Prostacyclin	Tissue factor
EDRF	PAF
t-PA, u-PA (plasminogen activator of tissue or urokinase type)	Factor V
	Factor VIII
Heparan sulfate proteoglycans	Factor IX receptor
Thrombomodulin	Factor X receptor
Protein S	PAI-1

agulant functions of EC. Tissue factor production plays a central role in this alteration [8]. That such a disturbance will promote the development of microthrombi is evident. Numerous inflammatory mediators, including the cytokines, are able to up-regulate the prothrombogenic activity of the endothelium [9].

Endothelium and Vascular Tone

Intensive care physicians have known the characteristics of peripheral vascular insufficiency for a long time. From a pathophysiological viewpoint, the striking feature is the altered function of smooth muscle activity in the microcirculation. EC appear to be instrumental in these changes on the basis of their production of vasodilating and vasoconstricting factors. The most important of the former are prostacyclin and NO (= EDRF, endothelium-derived relaxing factor), whilst the essential vasoconstrictor is the potent endothelin-1 (ET-1). One of the well-known aspects of septic shock is the marked reduction of peripheral vascular tone, implicating a predominance of vasodilatory activity, although it must be taken into consideration that certain elements of the microcirculation can show increased smooth muscle activity. Thus, in the early phase of septic shock, peripheral *venous* tone increases [10]. Furthermore, it is evident that, for example, endotoxin can alter the sensitivity of arteriolar smooth muscle to vasoactive signals such as norepinephrine and vasopressin [11]. A further significant aspect is the heterogeneity of the microcirculation, so that the response to a stimulus in one vascular bed may be entirely different from that in another. In this respect the reduction of endothelial NO release in the splanchnic circulation as a result of ischemia and reperfusion requires more attention [12]. A delineation of the exact roles of EC factors is complicated by the fact that numerous other vasodilatory and vasoconstricting substances are produced by other members of the inflammatory orchestra, including blood cells and plasma mediator systems. Other vasodilators include kinins, serotonin, histamine and eicosanoids.

Endothelium and Amplification of Inflammation

This very complex field of the shifting of a controlled inflammatory process to that of a self-perpetuating amplification of inflammation involves all major components of inflammation: mediators (including bacterial toxins, cytokines, arachidonic acid metabolites, kinins, complement factors and products of the coagulation and fibrinolytic cascades), blood cells (especially monocytes, neutrophilic granulocytes and platelets) and EC. Focussing on the latter, three aspects should be emphasized.

(1) Autocrine loops for mediator synthesis: This interesting cell biological phenomenon of a cell producing a substance, stimulating further synthesis in the same cell was shown as early as 1987 for IL-1 (interleukin-1β) [13]. Thus, IL-1 can induce the synthesis of IL-1 in EC.

(2) Priming effects of mediators: Priming action involves the ability of a biologically insignificant concentration of a mediator to prime a biological system, such as a cell, to respond dramatically to a second mediator in a concentration of the latter which alone evokes no significant or only minimal activity. Relevant priming agents with regard to the pathogenesis of MODS appear to be platelet-activating factor (PAF; 1-O-alkyl-2(R)-acetylglycero-3-phosphocholine), which is produced by a variety of cells, including EC, monocytes and granulocytes [14], as well as other cytokines [15]. Braquet et al. [16] cite various examples of how PAF, TNF-α, IL-1 and GM-CSF interact with EC and blood cells in ways in which one of the cytokines acts at very low concentration as a primer. Thus, for example, TNF-α can prime human neutrophilic granulocytes to respond to PAF stimulation by a marked increase in superoxide anion release, a factor with damaging effects on EC.

(3) Up-regulation of CAM expression: The regulation of the expression of both consitutive and inducible cell adhesion molecules (CAM) on the luminal surface of EC represents an integral part of the physiological response to injury by controlling the interaction between blood cells and the endothelium. These CAM belong to two molecular families: the selectins, represented by E-selectin and P-selectin (CD62) and the immunoglobulin gene superfamily, represented by ICAM-1, ICAM-2 and VCAM-1 [17]. Important mediators produced in MODS, such as endotoxin, TNF-α and IL-1, can elicit a marked up-regulation of CAM on the EC surface [18]. This phenomenon can be well demonstrated on EC in vitro (fig. 1).

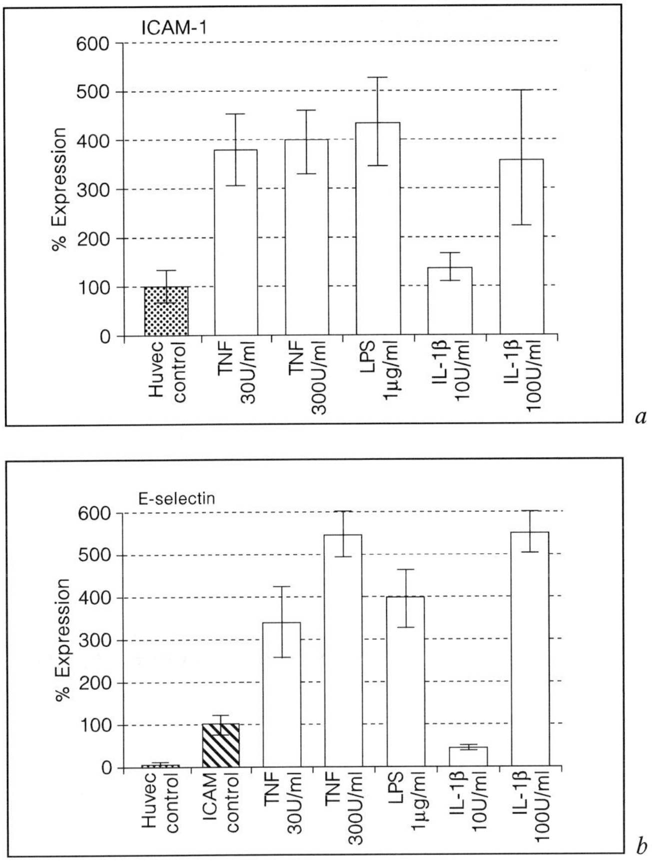

Fig. 1. a, b Up-regulation of ICAM-1 and E-selectin on cultured human umbilical vein endothelial cells (HUVEC) in passage 2. In both histograms the constitutive expression of ICAM-1 on unstimulated cells was set at 100% and all other values given as a relative percentage. The assay involved spectrophotometric measurement in a cell EIA (enzyme immunoassay). In both *a* and *b* it can be seen that TNF-α (30 and 300 U/ml), LPS (1 µg/ml) and IL-1β (10 and 100 U/ml) all elicit a statistically significant ($p < 0.001$) increase in CAM expression compared to unstimulated controls. Other experiments showed that LPS concentrations of as low as 1 ng/ml were able to stimulate CAM expression on HUVEC.

That these considerations are not merely academic has been well demonstrated by a baboon shock model. Redl et al. [19] studied the expression of E-selectin in the vascular system of the baboon under hypovolemic shock or in septic shock, the latter induced by intravenous injection of live *Escherichia coli* bacteria. Whilst in the former there was only minimal expression of E-selectin, the septic shock model gave marked expression in various vascular beds.

Endothelium and Growth-Regulating Signals

In the wall of blood vessels there is a strict growth-regulatory control exerted by both EC and smooth muscle cells (SMC). In this physiological situation, both cell types synthesize growth-inhibitory substances. However, under certain stimuli very potent growth-regulating molecules can be produced, including transforming growth factor-beta-1 (TGF-β_1), platelet-derived growth factor (PDGF), basic fibroblast growth factor (b-FGF), IL-1, insulin-like growth factor I (IGF-I) and various proteoglycan species. Cytokine signals appear to be able to up-regulate the synthesis of some of these mitogenic substances. Thus, for example, TNF-α can stimulate EC to produce PDGF [20]. In vitro studies indicate that neutrophilic granulocytes (PMN) induce the release of PDGF and b-FGF from EC [21]. The significance of this for MODS becomes apparent when one considers that ARDS, one of the hallmarks of MODS, involves fibroblastic proliferation and increased matrix synthesis in the lung.

The Role of Bacterial Cell Wall Products

The close association of sepsis and MODS is well established [22]. This has led to intensive studies on the significance of substances produced by microorganisms, and, in particular, bacteria. Although much attention has been focused on endotoxins (lipopolysaccharides, LPS) from gram-negative bacteria [23], there is accumulating evidence that peptidoglycans from gram-positive bacteria are also able to elicit critical pathogenetic reactions in the host organism [24]. Concerning LPS, it is now clear that LPS binds with high affinity via its lipid A moiety to a serum protein, LPS-binding protein (LBP). The LPS-LBP complexes then interact with membrane-anchored CD14, especially on blood monocytes, to induce activation, with subsequent production of cytokines, such as TNF-α and IL-1 [25, 26]. With respect to the endothelium, LPS exerts the following important effects:

(1) Direct damage to EC: Using cultivated pulmonary artery EC, Meyrick et al. [27] were able to show that *E. coli* endotoxin at concentrations as low as 1 ng/ml could elicit structural and functional alterations.

(2) Activation of mediator systems: This applies not only to cell-associated mediator cascades, but also to mediator systems in plasma. An essential step in activation of the latter is the activation of coagulation factor XII (Hagemann factor), which can be achieved both by LPS and by peptidoglycans [24]. LPS can also induce EC production of tissue factor [28] and von Willebrand factor [29]. Further significant actions of LPS are the suppression of EC thrombomodulin [30] and the induction of plasminogen activator inhibitor-1 [31, 32].

(3) Up-regulation of CAM expression: The importance of this phenomenon has been stressed above. It should, however, be added that various CAM are involved at different times following stimulation. Thus, for example, P-selectin expression appears to play a vital role in the first few minutes after LPS stimulation and is responsible along with PAF for early neutropenia [33], whereas E-selectin reaches a maximal expression after 4–6 h [34]. On the other hand, ICAM-1 requires 8–12 h to peak and can remain at elevated levels of expression for more than 48 h [18, 35].

Problems and Perspectives

Although there is accumulating evidence from in vitro and in vivo studies that the endothelium plays a central role in the pathogenesis of MODS, there are still some essential questions which remain unanswered. These include the following:

(1) Do systemic mediator levels reflect the local status? This important issue concerns not just better understanding of the pathogenesis, but also prognostic aspects.

(2) Which role does the heterogeneity of the microcirculation play? As corollary of our hypothesis that the endothelium is central to the development of MODS, the role of organ-specific effects of the endothelium needs to be considered. In this context the barrier function of the hepatic endothelium against intestinal bacterial toxins requires more intensive investigation, although there are indications that sinusoidal EC may be less easily stimulated by cytokines to increase procoagulant activity [36].

(3) What are the key events in endothelial dysfunction and what is their temporal sequence?

These questions might appear *prima vista* to represent an academic exercise. However, the principle of targeted intervention has as a prerequisite the detailed knowledge of the pathogenetic events. The application of monoclonal antibodies to the treatment of MODS has already reached an advanced stage, although the initial promise of success has only been partially fulfilled by the conducted clinical trials [37]. Of possible future significance is the use of genetic manipulation, for example by introducing antisense m-RNA in order to regulate certain processes [38]. An example of such manipulation in an in vitro system was given by Itoh et al. [39], who introduced an antisense oligonucleotide complementary to b-FGF mRNA into EC in vitro. As a result, b-FGF production was reduced, with concomitant suppression of DNA synthesis. These experiments indicate that it is possible to specifically interfere with cellular synthetic activity, although numerous technical, not to mention ethical difficulties still need to be overcome.

In conclusion, it may be stated that the endothelium has a polyvalent role to play in the pathogenesis of MODS and that many of the essential elements in this role offer the possibility of specific therapeutic intervention. Whilst there is no reason for euphoria, there is legitimate cause for hope that certain vital pathogenetic steps can be effectively regulated. This clearly remains a challenge for both basic sciences and clinical medicine.

References

1 Bone RC, Balk RA, Cerra FB, Dellinger RP, Fein AM, Knaus WA, Schein RMH, Sibbald WJ: Definitions for sepsis and organ failure and guidelines for the use of innovative therapies in sepsis. Chest 1992;101:1644–1655.

2 Kirkpatrick CJ: Aspekte der Pathogenese des Multiorganversagens; in Vogt S, Metzner G, Haverich A (eds): Immundiagnostik und Immuntherapie in der modernen Intensivmedizin. Langerich, Pabst, 1994, pp 99–119.

3 Mittermayer Ch, Riede UN, Bleyl U, Herzog H, v Wichert P, Riesner K: Schocklunge. Verh Dtsch Ges Pathol 1978;62:11–65.

4 Baue AE: The horror autotoxicus and multiple-organ failure. Arch Surg 1992;127:1451–1462.

5 Bone RC: Modulators of coagulation. A critical appraisal of their role in sepsis. Arch Intern Med 1992;152:1381–1389.

6 Czervionke RL, Smith JB, Hoak JC, Fry GL, Haycraft DL: Use of a radioimmunoassay to study thrombin-induced release of PGI_2 from cultured endothelium. Thromb Res 1979;14:781–786.

7 Radomski MW, Palmer RMJ, Moncada S: The anti-aggregating properties of vascular endothelium: Interactions between prostacyclin and NO. Br J Pharmacol 1987;92:639–646.

8 Ruf W, Edgington TS: Structural biology of tissue factor, the initiator of thrombogenesis in vivo. FASEB J 1994;8:385–390.

9 Billiau A, Van de Kerckhove F: Cytokines and their interactions with other inflammatory mediators in the pathogenesis of sepsis and septic shock. Eur J Clin Invest 1991;21:559–573.

10 Astiz ME, Tilly E, Rackow ED, Weil MH: Peripheral vascular tone in sepsis. Chest 1991;99:1072–1075.

11 Baker CH, Sutton ET, Dietz JR: Endotoxin alteration of muscle microvascular renin-angiotensin responses. Circ Shock 1992;36:224–230.

12 Carey C, Siegfried MR, Ma X-L, Weyrich AS, Lefer AM: Antishock and endothelial protective actions of a NO donor in mesenteric ischemia and reperfusion. Circ Shock 1992;38:209–216.

13 Warner SJC, Anger KR, Libby P: Interleukin-1 induces interleukin-1. II. Recombinant human interleukin-1 induces interleukin production by adult vascular endothelial cells. J Immunol 1987; 139:1911–1917.

14 Braquet P, Touqui L, Shen TS, Vargaftig BB: Perspectives in platelet-activating factor research. Pharmacol Rev. 1987;39:97–145.

15 Koltai M, Hosford D, Braquet P: Role of PAF and cytokines in microvascular tissue injury. J Lab Clin Med 1992;119:461–466.

16 Braquet P, Paubert-Braquet M, Bourgain RH, Bussolino F, Hosford D: PAF/cytokine auto-generated feedback networks in microvascular immune injury: Consequences in shock, ischemia and graft rejection. J Lipid Mediat 1989;1:75–112.

17 Bevilacqua MP, Nelson RM: Endothelial-leukocyte adhesion molecules in inflammation and metastasis. Thromb Haemost 1993;70:152–154.

18 Albelda SM, Smith CW, Ward PA: Adhesion molecules and inflammatory injury. FASEB J 1994;8: 504–512.

19 Redl H, Dinges HP, Buurman WA, van der Linden CJ, Pober JS, Cotran RS, Schlag G: Expression of endothelial leukocyte adhesion molecule-1 in septic but not traumatic/hypovolemic shock in the baboon. Am J Pathol 1991;139:461–466.

20 Hajjar KA, Hajjar DP, Silverstein RL, Nachman RL: Tumor necrosis factor-mediated release of platelet-derived growth factor from cultured endothelial cells. J Exp Med 1987;166:235–245.
21 Totani L, Piccoli A, Pellegrini G, Di Santo A, Lorenzet R: Polymorphonuclear leukocytes enhance release of growth factors by cultured endothelial cells. Arterioscler Thromb 1994;14:125–132.
22 Bell RC, Coalson JJ, Smith JP, Johanson WG Jr: Multiple system organ failure and infection in adult respiratory distress syndrome. Ann Intern Med 1983;99:293–298.
23 Raetz CRH: Bacterial endotoxins: Extraordinary lipids that activate eucaryotic signal transduction. J Bacteriol 1993;175:5745–5753.
24 Glauser M, Zanetti G, Baumgartner J, Cohen J: Septic shock: Pathogenesis. Lancet 1991;338:732–736.
25 Tracey KJ: Tumor necrosis factor (cachectin) in the biology of septic shock syndrome. Circ Shock 1991;35:123–128.
26 Pugin J, Ulevitch RJ, Tobias PS: A critical role for monocytes and CD14 in endotoxin-induced endothelial cell activation. J Exp Med 1993;178:2193–2200.
27 Meyrick BO, Ryan US, Brigham KL: Direct effects of *E. coli* endotoxin on structure and permeability of pulmonary endothelial monolayers and the endothelial layer of intimal explants. Am J Pathol 1986;122:140–151.
28 Moldow CF, Bach RR, Staskus K, Rick PD: Induction of endothelial tissue factor by endotoxin and its precursors. Thromb Haemost 1993;70:702:706.
29 Gralnick HR, McKeown LP, Wilson OM, Williams SB, Elin RJ: Von Willebrand factor release induced by endotoxin. J Lab Clin Med 1989;113:118–122.
30 Moore RL, Andreoli SP, Esmon NL, Esmon CT, Bang ND: Endotoxin enhances tissue factor and suppresses thrombomodulin expression of human vascular endothelium in vitro. J Clin Invest 1987;79:124–130.
31 Van den Berg EA, Sprengers ED, Burgess JM, Maciag T, van Hinsbergh VWM: Regulation of plasminogen activator inhibitor-1 mRNA in human endothelial cells. Thromb Haemost 1988;60:63–67.
32 Colucci M, Paramo JA, Collen D: Generation in plasma of the fast-acting inhibitor of plasminogen activation in response to endotoxin stimulation. J Clin Invest 1985;75:818–824.
33 Coughlan AF, Hau H, Dunlop LC, Berndt MC, Hancock WW: P-selectin and platelet-activating factor mediate initial endotoxin-induced neutropenia. J Exp Med 1994;179:329–334.
34 Bevilacqua MP, Pober JS, Mendrick DL, Cotran RS, Gimbrone MA: Identification of an inducible endothelial-leukocyte adhesion molecule. Proc Natl Acad Sci USA 1987;84:9238–9242.
35 Rothlein R, Czajkowski M, O'Neill MM, Marlin S, Mainolfi E, Merluzzi VR: Induction of intercellular adhesion molecule 1 on primary and continuous cell lines by proinflammatory cytokines. J Immunol 1988;141:1665–1669.
36 Rieder H, Meyer zum Büschenfelde K-H, Ramadori G: Functional spectrum of sinusoidal endothelial liver cells. Filtration, endocytosis, synthetic capacities and intercellular communication. J Hepatol 1992;15:237–250.
37 Manthous CA, Hall JB, Samsel RW: Endotoxin in human disease. 2. Biologic effects and clinical evaluations of anti-endotoxin therapies. Chest 1993;104:1872–1881.
38 Brigham KL: Oxygen radicals – An important mediator of sepsis and septic shock. Klin Wochenschr 1991;69:1004–1008.
39 Itoh H, Mukoyama M, Pratt RE, Dzau VJ: Specific blockade of basic fibroblast growth factor gene expression in endothelial cells by antisense oligonucleotide. Biochem Biophys Res Commun 1992;188:1205–1213.

Prof. C.J. Kirkpatrick, MD, PhD, DSc, Institut für Pathologie, Klinikum der Johannes-Gutenberg-Universität, Langenbeckstrasse 1, D-55101 Mainz (Germany)

Sieberth HG, Stummvoll HK, Kierdorf H (eds): Continuous Extracorporeal Treatment in Multiple Organ Dysfunction Syndrome. Contrib Nephrol. Basel, Karger, 1995, vol 116, pp 10–15

Mediators in Continuous Extracorporeal Treatment of Multiple Organ Dysfunction Syndrome

G. Sunder-Plassmann, W.H. Hörl

Division of Nephrology, Department of Internal Medicine, University of Vienna, Austria

Mediators in the pathogenesis of sepsis include tumor necrosis factor-α (TNF-α), interleukins (ILs), platelet-activating factor (PAF), leukotrienes, thromboxane A_2, and activators of the complement cascade [1]. Furthermore, adhesion molecules, kinins, thrombin, myocardial depressant substances, β-endorphin, and heat-shock proteins participate in the development of sepsis [1].

Cytokine Levels in Systemic Inflammatory Response Syndrome (SIRS)

A decrease in the time course of TNF-α serum levels was associated with a favorable outcome of sepsis patients [2]. However, there was no difference of IL-6 levels between survivors and nonsurvivors of sepsis syndrome [2]. PAF has been discussed to be of predictive value in patients with sepsis. The half-life of PAF was found to be prolonged in patients with worst outcome of sepsis due to depression of acetylhydrolase activity [3]. High plasma concentrations of soluble E-selectin have been shown to be closely associated with multiple organ dysfunction and death [4]. In contrast, Donnelly et al. [5] found plasma concentrations of cytokines (TNF-α, IL-1, IL-6, IL-8), endotoxin or complement fractions of no prognostic value concerning the development of adult respiratory distress syndrome following severe accidental injury.

Froon et al. [6] found significantly higher plasma peak concentrations of the soluble 55,000-dalton molecular weight TNF receptor in nonsurvivors compared with survivors of sepsis syndrome. The difference in TNF peak concentrations and in the soluble 75,000-dalton TNF receptor concentrations between survivors and nonsurvivors was not significant in contrast to IL-6 peak concentrations.

Since plasma-soluble TNF receptor concentrations correlated with plasma creatinine concentration, it was concluded that this increase in patients with sepsis syndrome is merely the result of renal failure complicating sepsis [6].

Marty et al. [7] observed a significant difference between the concentration of IL-8 in sepsis survivors and nonsurvivors (81 ± 13 vs. 3,326 ± 1,219 pg/ml). However, although the mortality rate of nonseptic, multiple organ failure patients was very high (92%), low plasma concentration of IL-8 (78 ± 34 pg/ml) was found by the authors. These data clearly indicate that low cytokine plasma levels are not necessarily associated with favorable outcome of critically ill patients.

The recent finding of Puguin et al. [8, 9], that CD14-LPS complexes stimulate endothelial cells to release large amounts of cytokines, provides hope for future therapeutic strategies in patients with gram-negative sepsis to overcome the SIRS.

Pharmacological Antagonism of Inflammatory Mediators

Cytokines play a major role in the response to injury and infection and the development of organ damage in critically ill patients [10]. TNF antagonists or antibodies and IL-1 receptor blockers have been shown to reduce mortality in patients and animals with septic shock [11–16]. Initial evaluation of recombinant human IL-1 receptor antagonist in the treatment of sepsis syndrome demonstrated that the survival advantage increases with increasing severity of illness [17]. However, final analysis of 893 cases revealed no statistical significant increase of survival in patients with sepsis syndrome treated with the recombinant human IL-1 receptor antagonist [18]. According to Tompkins [19], the Gordian knot was not unraveled by the blockade with IL-1 receptor antagonist in the systemic inflammatory response syndrome. Furthermore, adequate survival models and severity assessment methods for therapeutic evaluation are an important factor [20].

Extracorporeal Elimination of Inflammatory Mediators

Alternatively, the relative small molecular weight of the cytokines should allow their clearance during continuous hemodiafiltration (CVVHDF). Thus, newer indications for continuous renal replacement therapies should be the removal of chemical mediators such as PAF, IL-1 and TNF-α [21]. Storck et al. [22] found a significantly higher survival rate in patients with acute renal failure after surgery treated with pump-driven hemofiltration than with spontaneous continuous hemofiltration (CAVH) (29.4 vs. 12.5%). It was suggested that the better survival rate in the pump-driven hemofiltration group with 15.7 liters/day (CAVH group 7.5 liters/day) was due to faster elimination of toxic mediators [22].

Several studies demonstrate that continuous hemofiltration improves cardiac and lung function in patients with septic, posttraumatic, and post-operative shock [23–30]. These effects may not be related to fluid removal but to elimination of toxic mediators. Therefore, Lee et al. [31] performed a prospective, randomized controlled animal study with age-matched controls in order to evaluate the efficacy of pump-assisted continuous hemofiltration in lethal *Staphylococcus aureus*-induced sepsis in immature swine. Under three different conditions, filtered animals survived significantly longer than their matched controls. Percent increment values of survival time for filtered versus nonfiltered animals ranged from 57 to 312%.

The detection of TNF-α in ultrafiltrates of patients with sepsis prompted Bellomo et al. [32, 33] to investigate the clearance and removal rates of TNF-α and IL-6 as well as serum concentrations in septic patients treated with CVVHDF. The authors concluded that relevant amounts of cytokines are removed from the circulation of septic patients and that this phenomenon may prove to be of benefit for patients with sepsis [32, 33]. Elliott et al. [34] studied the effluent from patients on continuous arteriovenous hemofiltration with dialysis (CAVH-D) or continuous venovenous hemofiltration with dialysis (CVVH-D) for TNF-α and IL-1. Daily TNF-α removal was 88.4 ± 59 μg, daily IL-1 removal was 70.9 ± 171 μg. Total daily effluent volume was 35.6 ± 4.5 liters [34]. In a similar manner, Millar et al. [35] found a reduction of TNF-α but not of IL-6 or IL-8 serum levels in children with multiorgan dysfunction following cardiopulmonary bypass treated with hemofiltration. In contrast, Byrick et al. [36] did not observe a decreased TNF-α concentration in severe rhabdomyolysis treated by CAVH-D. Similarly, Kierdorf [37] found no reduction of increased TNF-α values in nonseptic and septic patients with multiple organ failure despite removal of 5,000 ng TNF-α in nonseptic and 13,000 ng TNF-α in septic patients per week treated with CVVH. In the study of Andreasson et al. [38], complement levels increased during cardiopulmonary bypass and could be reduced by hemofiltration.

However, the demonstration of the presence of mediators in the ultrafiltrate does not necessarily indicate clinical importance. It has been argued that cytokines are middle molecular weight substances (e.g. molecular weight of IL-1β is 17,500 daltons, of TNF-α monomer is 17,000 daltons) which limits the rate of diffusion through the dialysis membrane [39]. Moreover, the biologically active form of TNF-α is known to be a trimer, with a molecular weight of 54,000 daltons [40]. Furthermore, cytokine passage through the membrane is limited by their binding on e.g. α_2-macroglobulin (molecular weight of 720,000 daltons) or soluble receptors, which makes the beneficial effect of cytokine removal questionable [39].

A well-known side effect of extracorporeal circulation of blood is a decrease of body temperature. Böhrer et al. [41] mentioned that restored normothermia due to blood cooling in the extracorporeal tubing, and not only due to removal of cytokines, may contribute to the improved outcome in their patients who received early CVVH for SIRS without major disturbances of body function.

The effect of low- and high-volume hemofiltration outcome of ICU patients or septic animals was recently summarized by Grootendorst and van Bommel [42]. In animal studies, hemofiltration improved hemodynamics and gas exchange in septic or endotoxin shock. High-volume convective blood purification provided a more beneficial hemodynamic effect as compared to low-volume hemofiltration. In clinical studies, an improvement of gas exchange and hemodynamics was observed in critically ill patients during hemofiltration due to fluid removal and convective elimination of mediators known to have vasoactive and/or cardiodepressant properties [42].

Schiffl et al. [43] demonstrated improved survival of ICU patients dialyzed with a biocompatible high-flux dialyzer compared to ICU patients treated with conventional cuprophane hemodialysis due to marked reduction of infectious complications. Rats exposed to cuprophane displayed a delayed recovery from acute renal failure and revealed an increase in activated neutrophils in the renal parenchyma [44]. It has been shown that neutrophils activated during sepsis are further activated during hemodialysis with cuprophane dialyzers [45]. Furthermore, granulocyte inhibitory proteins can be removed using high-flux dialyzer therapy [46].

Conclusion

There is no doubt that several inflammatory mediators and granulocyte inhibitory proteins as well as vasoactive and/or cardiodepressant substances can be measured in high-flux dialyzer ultrafiltrates of ICU patients. However, it is still unclear whether the clearances of the available filters are high enough to remove adequate amounts of these mediators (substances) to improve markedly the outcome of ICU patients by continuous extracorporeal treatment.

References

1 Bone RC: The pathogenesis of sepsis. Ann Intern Med 1991;115:457–469.
2 Martin C, Saux P, Mege JL, Perrin G, Papazian L, Gouin F: Prognostic values of serum cytokines in septic shock. Intensive Care Med 1994;20:272–277.
3 Graham RM, Stephens CJ, Silvester W, Leong LLL, Sturm MJ, Taylor RR: Plasdma degradation of platelet-activating factor in severely ill patients with clinical sepsis. Crit Care Med 1994;2:204–212.

4 Cowley HC, Heney D, Gearing AJ, Hemingway I, Webster NR: Increased circulating adhesion molecule concentrations in patients with the systemic inflammatory response syndrome: A prospective cohort study. Crit Care Med 1994;22:651–657.
5 Donnelly TJ, Meade P, Jagels M, Gryer G, Law MM, Hugli TE, Shoemaker WC, Abraham E: Cytokine, complement, and endotoxin profiles associated with the development of the adult respiratory distress syndrome after severe injury. Crit Care Med 1994;5:768–776.
6 Froon AHM, Bemelmans MHA, Greve JW, van der Linden CJ, Buurman WA: Increased plasma concentrations of soluble tumor necrosis factor receptors in sepsis syndrome: Correlation with plasma creatinine values. Crit Care Med 1994;22:803–809.
7 Marty C, Misset B, Tamion F, Fitting C, Carlet J, Cavaillon JM: Circulating interleukin-8 concentrations in patients with multiple organ failure of septic and nonseptic origin. Crit Care Med 1994;4: 673–679.
8 Pugin J: Bacteremia, sepsis and shock. Intensive Care Med 1994;20:92–93.
9 Pugin J, Ulevitch RJ, Tobias PS: A critical role for monocytes and CD14 in endotoxin-induced activation of endothelial cells. J Exp Med 1993;178:2193–2200.
10 Bellomo R: The cytokine network in the critically ill. Anaesth Intens Care 1992;3:288–302.
11 Marcecaux G, Vincent JL, Kentos A: Anti-TNF antibodies administration in human septic shock: A pilot study (abstract). Intensive Care Med 1990;16:S102.
12 Lilly CM, Sandhu JS, Ishizaka A, Harada H, Yonemaru M, Larrick JW, Shi TX, O'Hanley PT, Raffin TA: Pentoxifylline prevents tumor necrosis factor-induced lung injury. Am Rev Respir Dis 1989;139:1361–1368.
13 Waxman K: Pentoxifylline in septic shock (editorial). Crit Care Med 1990;18:243–244.
14 Evans D, Jacobs D, Revhaug A, Wilmore D: The effects of tumor necrosis factor and their selective inhibition by ibuprofen. Ann Surg 1989;209:312–321.
15 Wakabayashi G, Gelfand JA, Burke JF, Thompson RA, Dinarello CA: A specific receptor antagonist for interleukin-1 prevents *Escherichia coli*-induced shock in rabbits. FASEB J 1991;5:338–343.
16 Ohlsson K, Bjork P, Bergenfeldt M, Hageman R, Thompson RC: Intereukin-1 receptor antagonist reduces mortality from endotoxin shock. Nature 1990;348:550–552.
17 Fisher CJ, Slotman GJ, Opal SM, et al: Initial evaluation of human recombinant interleukin-1 receptor antagonist in the treatment of sepsis syndrome: A randomized, open-label, placebo-controlled multicenter trial. Crit Care Med 1994;22:12–21.
18 Fisher CJ, Dhainaut JFA, Opal SM, Pribble JP, Balk RA, Slotman GJ, Iberti TJ, Rackow EC, Shapiro MJ, Greenman RL, Reines HD, Shelly MP, Thompson BW, LaBrecque JF, Catalano MA, Knaus WA, Sadoff JC: Recombinant human interleukin-1 receptor antagonist in the treatment of patients with sepsis syndrome. JAMA 1994;23:1836–1843.
19 Tompkins RG: Human recombinant interleukin-1 receptor antagonist in the treatment of sepsis syndrome. Crit Care Med 1994;22:1361–1368.
20 Knaus WA, Harell FE, Fisher CJ, Wagner DP, Opal SM, Sadoff JC, Draper EA, Walawander CA, Conboy K, Grasela TH: The clinical evaluation of new drugs for sepsis: A prospective study design based on survival analysis. JAMA 1993;10:1233–1241.
21 Ronco C: Continuous renal replacement therapies for the treatment of acute renal failure in intensive care patients. Clin Nephrol 1993;3:187–198.
22 Storck M, Hartl WH, Zimmerer E, Inthorn D: Comparison of pump-driven and spontaneous continuous haemofiltration in postoperative acute renal failure. Lancet 1991;337:452–455.
23 Gotloib L, Barilay E, Shustak A, Lev A: Sequential hemofiltration in nonoliguric high capillary permeability pulmonary edema of severe sepsis: Preliminary report. Crit Care Med 1984;12:997–1000.
24 Coraim F, Wolner E: Management of cardiac surgery patients with continuous arteriovenous hemofiltration; in Sieberth HG, Mann H (eds): Continuous Arteriovenous Hemofiltration. International Conference on CAVH Aachen, Nov 1984. Basel, Karger, 1985, pp 103–110.
25 Coraim F, Fasol R, Stellwag F, et al: Continuous arteriovenous hemofiltration after cardiac surgery; in Sieberth HG, Mann H (eds): Continuous Arteriovenous Hemofiltration. International Conference on CAVH Aachen, Nov 1984. Basel, Karger, 1985, pp 116–124.

26 Ossenkoppele GJ, van der Meulen J, Bronsveld W, Thus LG: Continuous arteriovenous hemofiltration as an adjunctive therapy for septic shock. Crit Care Med 1985;13:102–104.

27 Radzivil GG, Beloborodov VB, Broniakin IV, et al: Hemosorption as a method of treating endotoxicosis in patients with septic shock of meningococcal etiology. Anesteziol Reanimatol 1987;2:46–50.

28 Gotloib L, Barzilay E, Shustak A, Wais Z, Jaichenko J, Lev A: Hemofiltration in septic ARDS. The artificial kidney as the artificial endocrine lung. Resuscitation 1986;13:123–132.

29 Zobel G, Trop M, Ring E, Grubbauer HM: Arteriovenous hemofiltration in children with multiple organ system failure. Int J Artif Organs 1987;10:233–238.

30 Barzilay E, Kessler D, Berlot G, Gullo A, Geber D, Zeev IB: Use of extracorporeal supportive techniques as additional treatment for septic-induced multiple organ failure patients. Crit Care Med 1989;17:634–637.

31 Lee PA, Matson JR, Pryor RW, Hinshaw LB: Continuous arteriovenous hemofiltration therapy for *Staphylococcus aureus*-induced septicemia in immature swine. Crit Care Med 1993;6:914–924.

32 Bellomo R, Tipping P, Boyce N: Tumor necrosis factor clearances during veno-venous hemodiafiltration in the critically ill. ASAIO Trans 1991;27:M322–M323.

33 Bellomo R, Tipping P, Boyce N: Continuous veno-venous hemofiltration with dialysis removes cytokines from the circulation of septic patients. Crit Care Med 1993;4:522–526.

34 Elliott D, Wiles CE, Reynolds HN: Removal of cytokines in septic patients using continuous veno-venous hemodiafiltration (letter). Crit Care Med 1994;22:718–719.

35 Millar AB, Armstrong L, van der Linden J, Moat N, Ekroth R, Westwick J, Scallan M, Lincoln C: Cytokine production and hemofiltration in children undergoing cardiopulmonary bypass. Ann Thorac Surg 1993;56:1499–1502.

36 Byrick RJ, Goldstein MB, Wong PY: Increased plasma tumor necrosis factor concentration in severe rhabdomyolysis is not reduced by continuous arteriovenous hemodialysis. Crit Care Med 1992;20:1483–1486.

37 Kierdorf HR: Einfluss der kontinuierlichen Hämofiltration auf Proteinkatabolismus, Mediatorsubstanzen und Prognose des akuten Nierenversagens im Multiorganversagen; Habschr, Aachen 1993.

38 Andreasson S, Göthberg S, Berggren H, Bengtsson A, Eriksson E, Risberg B: Hemofiltration modifies complement activation after extracorporeal circulation in infants. Ann Thorac Surg 1993;56: 1515–1517.

39 Schetz M: Removal of cytokines in septic patients using continuous veno-venous hemodiafiltration (letter). Crit Care Med 1994;4:715–716.

40 Beutler B: Endotoxin, tumor necrosis factor, and related mediators: New approaches to shock. New Horizons 1993;1:3–12.

41 Böhrer H, Schmidt H, Bach A, Motsch J, Martin E: Removal of cytokines in septic patients using continuous veno-venous hemodiafiltration (letter). Crit Care Med 1994;4:717–718.

42 Grootendorst AF, van Bommel EFH: The role of hemofiltration in the critically ill intensive care unit patient: Present and future. Blood Purif 1993;11:209–223.

43 Schiffl H, Lang SM, König A, Strasser T, Haider MC, Held E: Biocompatible membranes in acute renal failure: Prospective case-controlled study. Lancet 1994;344:570–572.

44 Schulman G, Fogo A, Gung A, Badr K, Hakim R: Complement activation retards resolution of ischemic renal failure in the rat. Kidney Int 1991;40:1069–1074.

45 Hörl WH, Schaefer RM, Hörl M, Heidland A: Neutrophil activation in acute renal failure and sepsis. Arch Surg 1990;125:651–654.

46 Haag-Weber M, Mai B, Cohen G, Hörl WH: GIP and DIP: A new view of uraemic toxicity. Nephrol Dial Transplant 1994;9:346–347.

Prof. Dr. Dr. W.H. Hörl, Medizinische Universitäts-Klinik III,
Klinische Abteilung für Nephrologie und Dialyse, Währinger Gürtel 18–20,
A–1090 Wien (Austria)

Sieberth HG, Stummvoll HK, Kierdorf H (eds): Continuous Extracorporeal Treatment in Multiple Organ Dysfunction Syndrome. Contrib Nephrol. Basel, Karger, 1995, vol 116, pp 16–21

Prognostic Importance of Vital Organ Dysfunction in Sepsis

Peter M. Suter[a], *Didier Pittet*[b]

[a] Division of Surgical Intensive Care,
[b] Infection Control, Division of Infectious Diseases, Hôpital Cantonal Universitaire, Geneva, Switzerland

Sepsis is emerging as a major cause of morbidity and mortality in the developed countries. In the United States, 500,000 patients develop sepsis each year, half of whom have associated positive blood cultures [1, 2]. It is estimated that about half the cases of sepsis with bacteremia show evidence of organ dysfunction [2]. The mortality associated with this syndrome averages 35%, ranging from 20 to 50%, and accounts for about 100,000 deaths/year, representing half the 200,000 deaths from all cases of sepsis [2, 3]. Nosocomial bacteremia is associated with higher mortality rates than community-acquired infection [4–7]. The underlying disease definitely contributes to the crude mortality from bloodstream infection [8, 9]. The attributable mortality defines the mortality rate directly attributable to the infection and distinguishes the contribution of the infection from those of the underlying medical conditions. The attributable mortality of nosocomial bloodstream infection has averaged 27% in six specifically designed studies, with point estimates ranging from 14 to 38% [3]. Anticipating an average attributable mortality of 27% for nosocomial bloodstream infections, we estimated that as many as 62,500 deaths directly attributable to the infection may occur each year in the United States [6]. In critically ill patients, we recently determined the attributable mortality of nosocomial bacteremia to be 35%, and the infection was associated with a significant prolongation of intensive care unit (ICU) and hospital stay [10].

Influence of Therapy

Appropriate antibiotic therapy influences the outcome [7, 9, 11]. However, despite the availability of potent antibiotics and modern intensive care, no change in overall fatality rate has been noted for over a decade. Clearly, there is a need for evaluation of novel therapeutic approaches. New adjunctive immunotherapies using antiendotoxin antibodies have been investigated for their potential benefit in gram-negative sepsis [12, 13]. However, they are apparently of marginal benefit with favorable outcome noted only in subsets of septic patients [12–14]. Similar therapeutic approaches are currently under investigation with monoclonal antibodies against tumor necrosis factor or with an interleukin-1 receptor antagonist. The latter therapies might be of benefit not only to patients with gram-negative infections, but also to those with gram-positive, fungal, or parasitic diseases.

Prognosis of Sepsis in the ICU

The incidence of sepsis is 2–5 times higher in ICUs than in other hospital wards [15, 16]. ICUs are also privileged investigational areas where the clinical course of sepsis can be easily observed because of the constant monitoring of vital organ functions. Dysfunction of vital organs is frequent in sepsis and is associated with a poor prognosis [12, 13, 17, 18]. Since some of the high-technology immunotherapies may prevent the development of organ dysfunction [12, 13, 19] and thereby improve outcome, valid predictive models of high-risk sepsis patients would help define candidates for these expensive agents. Very few studies address the specific question of the prognosis of sepsis in a cohort of ICU patients. Dominguez de Villota et al. [20] observed that, among patients admitted to medical ICU, those with septicemia had a 2-fold higher fatality rate than those without septicemia (62 vs. 28%, respectively). Among the 109 patients with septicemia, the mortality rate was twice as high in those developing septic shock compared to those who did not (91 vs. 49%, respectively). In a 5-year retrospective cohort study of all patients (n = 5,457) admitted in our surgical ICU [21], patients with bacteremic sepsis (n = 176) had a 5-fold increased risk of in-hospital death compared to nonsepticemic patients (RR 5.03, CI_{95} 4.2–6.1, $p < 0.0001$). This estimate persisted after stratification according to age, sex, primary diagnosis for and conditions of admission to the ICU.

Outside the intensive care setting, numerous studies attempted to identify prognostic factors in bacteremia [8, 9, 11, 22–24]. These works mainly emphasized the deleterious impact of advanced age, underlying diseases and septic shock on the outcome of bacteremia. Amongst others, factors frequently associated with higher mortality from bacteremia include certain classes of microorganism, certain anatomic sources, nosocomial acquisition of the infection, hypothermia, a low neutrophil count, and inappropriate antibiotic therapy.

Scoring Systems

Several severity-of-illness scoring systems have been proposed to estimate patient risk for mortality in ICUs objectively. Models for use in critically ill patients have been updated recently to improve accuracy in predicting survival [25–28]. Predicting mortality among critically ill patients with sepsis is, however, a difficult task, because sepsis is a complex, dynamic process. Knaus et al. [29] recently proposed a sepsis severity scoring system that modifies the Apache III score to estimate the 28-day mortality risk for patients with sepsis syndrome entering clinical trials. The model was derived using retrospective analysis of data collection from several large databases. The results of their investigation indicated that the acute physiological abnormalities are the most important prognostic factors influencing outcome. Additional weight assigned to some physiological variables beyond the weights accorded in the Apache III scoring system increased the explanatory power of 28-day mortality.

Importance of Comorbidities and Organ Failures

Unfortunately, the impact of comorbidities other than cirrhosis, dialysis and metastatic tumor, that might influence outcome, could not be studied in the model proposed by Knaus et al. [29] because most study entrance criteria excluded such patients. With the model by Knaus et al. [29] the Somers' Dyx and ROC values, for in-hospital as well as 21-day mortality, never exceeded 0.59 and 0.80, respectively, for 28-day mortality in patients with presumed gram-negative infection and sepsis syndrome.

The need for mechanical ventilatory support is an independent predictor of mortality of sepsis. Craven et al. [30] noted that acute respiratory failure occurring at any time during the hospital stay was an independent predictor of hospital fatality (RR = 2.5, CI_{95} 1.6–3.8, $p = 0.0001$).

The importance of multiple organ dysfunctions as a prognostic factor in sepsis has long been established [2, 12, 13, 17, 18]. Multiorgan dysfunction commonly affects kidneys, liver, lungs, heart, central nervous and coagulation systems. Organ dysfunction developing after the onset of sepsis markedly affects the outcome. Herbert et al. [31] showed a linear association between the number of organ system dysfunctions and 30-day mortality in a cohort of critically ill patients with sepsis syndrome. In their series, the authors described that patients with three or more organ system dysfunctions had a significantly ($p < 0.01$) higher mortality than did those with less than three, after adjustment for age and sex. The logistic model proposed by Herbert et al. [31] was however poorly sensitive (50.9%) in identifying nonsurvivors (to predict death).

Multiple organ dysfunction syndrome – formerly called multiple organ failure syndrome – is an intricate, dynamic, frequently progressive and ultimately fatal condition. The cause of this syndrome is not fully understood, but it currently accounts for most deaths in noncoronary ICUs in the United States. To our knowledge, the development of multiple organ dysfunction syndrome has never been studied systematically in a consecutive cohort of critically ill patients with bacteremic sepsis. Severity of multiple organ dysfunction syndrome is a function of the number and duration of the organ dysfunction(s) and is not associated with a specific organ dysfunction. Septic shock was long considered the motor of multiple organ dysfunction syndrome occurring in the setting of sepsis.

Fever and Microbiological Variables

Fever reflects the systemic response to the septic process. Weinstein et al. [7] observed an increased risk of death in bacteremic patients without fever. A negative correlation between the degree of febrile response at the onset of septicemia and outcome strongly suggests that the degree of the patient's response may influence the prognosis. Recent clinical studies demonstrated that hypothermia is associated with worse prognosis in patients with sepsis syndrome [32] and some authorities suggested including hypothermia in the list of potential prognostic factors in septic patients [33]. Agents that may alter the adequacy of fever response should be studied in clinical trials, since these agents could have deleterious effects.

Certain microbiological variables such as *Pseudomonas aeruginosa* and *Candida* spp. bloodstream infections seem important. Previous studies in different patient populations have suggested that these organisms are responsible for higher mortality [7, 12, 22, 24, 34, 35]. In these studies, however, outcome was not corrected for confounding factors such as severity of underlying diseases or evolving organ dysfunctions.

Any Therapeutic Implications?

Core endotoxin components of the outer membrane of gram-negative bacteria have been proved to be potent mediators of shock. Based on previously encouraging results that showed improved survival in patients with gram-negative bacteremia receiving polyclonal antibodies directed at the core structure [36], recent clinical trials were conducted in patients with different septic syndromes using monoclonal antiendotoxin antibodies [12, 13]. The conclusions of these studies have been debated. New molecules neutralizing key mediators of sepsis are currently undergoing clinical trials. One part of the analytic problem results from insufficient characterization of the risk factors that may influence prognosis. Better outcome predictions could improve risk stratification and the design of clinical trials.

Appropriately designed clinical trials should take into account both the physiological changes present at the onset of sepsis and those occurring during the subsequent clinical course of sepsis. Patient groups should be stratified at the onset of sepsis for the variables defining underlying comorbidities, organ dysfunction and physiologic derangements. Organ dysfunction developing after the onset of sepsis should be included in the final analysis. The effects of new therapies on the development of vital organ dysfunction and outcome could hopefully improve our understanding of the pathophysiologic mechanisms leading to death in septic ICU patients.

References

1 Center for Disease Control: Increase in national hospital discharge survey rates for septicemia – United States. MMWR 1990;39:31–34.

2 Bone RC, Fisher CJ JR, Clemmer TP, Slotman GJ, Metz CA, Balk R: Sepsis syndrome: A valid clinical entity. Methylprednisolone severe sepsis study group. Crit Care Med 1989;17:389–392.

3 Wenzel RP: The mortality of hospital-acquired bloodstream infections: Need for a new vital statistic? Int J Epidemiol 1988;17:225–227.

4 Bryan CS, Hornung CA, Reynolds KL, Brenner ER: Endemic bacteremia in Columbia, South Carolina. Am J Epidemiol 1986;123:113–127.

5 Filice GA, Van Etta LL, Darby CP, Fraser DW: Bacteremia in Charleston County, South Carolina. Am J Epidemiol 1986;123:128–136.

6 Pittet D: Nosocomial bloodstream infections; in Wenzel RP (ed): Prevention and Control of Nosocomial Infections, ed 2. Baltimore, Williams & Wilkins, 1993, p 512.

7 Weinstein MP, Murphy JR, Reller LB, Lichtenstein KA: The clinical significance of positive blood cultures: A comprehensive analysis of 500 episodes of bacteremia and fungemia in adults. II. Clinical observations, with special reference to factors influencing prognosis. Rev Infect Dis 1983;5:54–70.

8 McCabe WR, Jackson GG: Gram-negative bacteremia. I. Etiology and ecology. Arch Intern Med 1962;110:847–855.

9 Kreger BE, Craven DE, Carling PC, Mc Cabe WR: Gram-negative bacteremia. III. Reassessment of etiology, epidemiology and ecology in 612 patients. Am J Med 1980;68:332–343.

10 Pittet D, Tarara D, Wenzel RP: Nosocomial bloodstream infection in the critically ill patients: Excess length of stay, extra costs, and attributable mortality. JAMA 1994;271:1598–1601.

11 Bryan CS, Reynolds KL, Brenner ER: Analysis of 1,186 episodes of gram-negative bacteremia in non-university hospitals: The effects of antimicrobial therapy. Rev Infect Dis 1983;5:629–638.

12 Greenman RL, Schein RMH, Martin MA, Wenzel RP, Mac Intyre NR, Emmanuel G, Chmel H, Kohler RB, Mc Carthy M, Plouffe J, Russel JA: A controlled clinical trial of E5 murine monoclonal IgM antibody to endotoxin in the treatment of gram-negative sepsis. JAMA 1991;266:1097–1102.

13 Ziegler EJ, Fisher CJ Jr, Sprung CL, Straube RC, Sadoff JC, Foulke GE, Wortel CH, Fink MP, Dellinger RP, Teng NNH, Allen IE, Berger HJ, Knatterud GL, Lo Buglio AF, Smith CR: Treatment of gram-negative bacteremia and septic shock with HA-1A human monoclonal antibody against endotoxin. A randomized, double-blind, placebo-controlled trial. N Engl J Med 1991;324:429–436.

14 Wenzel RP: Anti-endotoxin monoclonal antibodies – A second look. N Engl J Med 1992;326:1151–1152.

15 Daschner FD, Frey P, Wolff G, Baumann PC, Suter PM: Nosocomial infections in intensive care wards: A multicenter prospective study. Intensive Care Med 1982;8:5–9.

16 Pittet D, Herwaldt LA, Massanari RM: The intensive care unit; in Brachman PS, Bennett JV (eds): Hospital Infections, ed 3. Boston, Little, Brown, 1992, p 405.

17 Bone RC, Fisher CJ Jr, Clemmer TP, Slotman GJ, Metz C: Early methylprednisolone treatment for septic syndrome and the adult respiratory distress syndrome (Erratum in Chest 1988;94:448). Chest 1987;92:1032–1036.
18 Veterans Administration Systemic Sepsis Cooperative Study Group: Effect of high-dose glucocorticoid therapy on mortality in patients with clinical signs of systemic shock. N Engl J Med 1987;317: 659–665.
19 Ohlsson K, Bjsrk P, Bergenfeldt M, Hageman R, Thompson RC: Interleukin-1 receptor antagonist reduces mortality from endotoxin shock. Nature 1990;348:550–552.
20 Dominguez de Villota E, Algora A, Rubio A, Roig M, Mosquera JM, Galdos P, Diez-Balda V: Septicemia in a medical intensive care unit. Clinical, biochemical and microbiological data of 109 cases. Intensive Care Med 1983;9:109–115.
21 Pittet D, Thiévent B, Wenzel RP, Li N, Gurman G, Suter PM: Importance of pre-existing comorbidities for prognosis of septicemia in critically ill patients. Intensive Care Med 1993;19:265–272.
22 Gatell JM, Trilla A, Latorre X, Almela M, Mensa J, Moreno A, Miro JM, Martinez JA, De Anta MTJ, Soriano E, Garcia San Miguel J: Nosocomial bacteremia in a large Spanish teaching hospital: Analysis of factors influencing prognosis. Rev Infect Dis 1988;10:203–210.
23 Pittet D, Li N, Wenzel RP: Association of secondary and polymicrobial nosocomial bloodstream infections with higher mortality. Eur J Clin Microbiol Infect Dis 1993;12:813–819.
24 Roberts FJ, Geere IW, Coldman A: A three-year study of positive blood cultures, with emphasis on prognosis. Rev Infect Dis 1991;13:34–46.
25 Knaus WA, Draper EA, Wagner DP, Zimmerman JE: Apache II: A severity of disease classification system. Crit Care Med 1985;13:818–829.
26 Knaus WA, Wagner DP, Draper EA, Zimmerman JE, Bergner M, Bastos PG, Sirio CA, Murphy DL, Lotring T, Damanio A, Harrell FE: The Apache III prognostic system. Risk prediction of hospital mortality for critically ill hospitalized adults. Chest 1991;100:1619–1636.
27 Le Gall JR, Lemeshow St, Saulnier F: A new simplified acute physiology score (SAPS II) based on a European/North American multicenter study. JAMA 1993;270:2957–2963.
28 Lemeshow S, Teres D, Klar J, Avrunin JS, Gehlbach SH, Rapoport J: Mortality probability models (MPM II) based on an international cohort of intensive care unit patients. JAMA 1993;270:2478–2486.
29 Knaus WA, Harrell FE, Fisher CJ, Wagner DP, Opal SM, Sadoff JC, Draper EA, Walawander CA, Conboy K, Grasela TH: The clinical evaluation of new drugs for sepsis. A prospective study design based on survival analysis. JAMA 1993;270;1233–1241.
30 Craven DE, Kunches LM, Lichtenberg DA: Nosocomial infection and fatality in medical surgical intensive care unit patients. Arch Intern Med 1988;148:1161–1168.
31 Herbert PC, Drummond AJ, Singer J, Bernard GR, Russell JA: A simple multiple system organ failure scoring system predicts mortality of patients who have sepsis syndrome. Chest 1993;104: 230–235.
32 Clemmer TP, Fisher CJ Jr, Bone RC, Slotman GJ, Metz CA, Thomas F: Hypothermia in the sepsis syndrome and clinical outcome. The Methylprednisolone Severe Sepsis Study Group. Crit Care Med 1992;20:1395–1401.
33 Balk RA, Parillo JE: Prognostic factors in sepsis: The cold facts. Crit Care Med 1992;20:1373–1374.
34 Bisbe J, Gatell JM, Puij J, Malloas J, Martinez JA, Jimenez-de-Anta MT, Soriano E: *Pseudomonas aeruginosa* bacteremia: Univariate and multivariate analysis of factors influencing the prognosis in 133 episodes. Rev Infect Dis 1988;10:629–635.
35 Miller PJ, Wenzer RP: Etiologic organisms as independent predictors of death and morbidity associated with bloodstream infections. J Infect Dis 1987;156:471–477.
36 Ziegler EJ, Mc Cutchan JA, Fierer J, Glauser MP, Sadoff JC, Douglas H, Braude AJ: Treatment of gram-negative bacteremia and shock with human antiserum to a mutant *Escherichia coli.* N Engl J Med 1982;307:1225–1230.

Peter M. Suter, MD, Division des Soins Intensifs Chirurgicaux,
Hôpital Cantonal Universitaire, CH–1211 Genève 14 (Switzerland)

Sieberth HG, Stummvoll HK, Kierdorf H (eds): Continuous Extracorporeal Treatment in Multiple Organ Dysfunction Syndrome. Contrib Nephrol. Basel, Karger, 1995, vol 116, pp 22–27

Intensive Care Management of Patients with Acute Hepatic and Renal Failure

Andrew Davenport[a], *Susan A. Kirby*[b]

[a] Renal Unit, Southmead Hospital, and
[b] Department of Anesthesia, Bristol Royal Infirmary, Bristol, UK

Despite a continuing high mortality of patients admitted with or who develop fulminant hepatic failure (FHF) [1], there has been some improvement in patient survival. This increased patient survival may be related to changes in intensive care management with careful monitoring of cardiovascular and cerebrovascular parameters becoming standard.

Some problems are directly associated with hepatic failure; coagulopathy and thrombocytopenia with an increased risk of gastrointestinal haemorrhage, endotoxaemia from gastrointestinal bacteria and reduced bacterial opsonization associated with deficiencies of both fibronectin and complement. In addition, patients with FHF develop a severe circulatory disturbance chatacterized by reduced systemic vascular resistance and an inadequate increase in cardiac index, resulting in cellular hypoxia [2]. In part, this excessive post-arteriolar vasodilation is due to the direct effect of circulating endotoxins acting on the endothelium increasing the production of nitric oxide and other potent vasodilators.

Acute oliguric renal failure commonly occurs in FHF following paracetamol (acetaminophen) self-poisoning, due to the toxic effect of the metabolite, N-acetyl-*p*-benzoquinoneimine, on the proximal convoluted tubules. The reported incidence of renal failure in cases of infective hepatitis varies between series [1]. Despite dialysis, patients who develop acute renal failure in association with acute or chronic liver disease have a greater mortality [1, 3]. These patients have more severe hepatic dysfunction, with greater mismatch of tissue oxygen demand/supply.

General Intensive Care Management Designed to Control Intracranial Pressure

Patients at risk of cerebral oedema should be nursed with a 10–20° head-up tilt to maximize cerebral perfusion pressure (CPP). Elective hyperventilation (arterial $PaCO_2$ between 3.5 and 4.5 kPa (26–34 mm Hg)) has not been proven to increase patient survival or prevent sudden surges in intracranial pressure (ICP) [4], but has been found to be effective in reducing ICP in patients with FHF complicated by acute oliguric renal failure [5], and is to be recommended. Patients are paralysed and sedated to facilitate tracheal toilette and physiotherapy without causing surges in ICP. We used continuous propofol infusions for sedation following reports of decreased ICP and reduced cerebral oxygen demand without adversely affecting cerebral perfusion pressure [6]. In addition, propofol may have an antioxidant effect.

As cardiovascular instability is a major problem, inotropic support with noradrenaline and prolonged infusion of *n*-acetyl-cysteine are currently used in many centres to maintain systolic blood pressure.

Mode of Renal Replacement Therapy

Patients with grade IV hepatic coma and established acute oliguric renal failure are at increased risk of dying from cerebra oedema/hypoxia [4]. Treatment with peritoneal dialysis failed to improve patient survival [3]. Although peritoneal dialysis may have some beneficial effects in removing gastrointestinal endotoxins by lavage, the disadvantages in terms of increased risk of infection, removal of proteins and difficulty in reliably maintaining fluid balance outweigh any possible advantages.

Intermittent haemodialysis and/or haemofiltration have been reported to cause an increase in ICP in association with a reduction in cerebral oxygen delivery and CPP [7]. ICP has been observed to increase both at the end of treatment, when the maximum gradient between plasma osmolality and intracerebral osmolality has developed, and also at the start of treatment [7]. These initial changes in ICP occur at a time when there are no detectable changes in plasma osmolality, sodium or pH. They are associated with cardiovascular changes; reduction in right- and left-sided cardiac filling pressures, MAP, increased pulmonary vascular resistance with no increase in systemic vascular resistance [11]. The net effect is a reduction in tissue oxygen delivery, exacerbating an already compromised cerebral neuronal oxygen supply, resulting in the development of vasogenic cerebral oedema and a further increase in ICP.

Continuous haemofiltration and/or dialysis has proved superior in maintaining cardiovascular and intracerebral stability in these patients [8]. Until recently, most continuous circuits have been spontaneously driven, and with relatively slow blood flows, changes in the 'effective' intravascular volume have been modest and correction made where appropriate, so accounting for the improved stability. At the greater blood flows achieved by pumped circuits, changes in the 'effective' intravascular volume occur more frequently and despite attempts to maintain isovolaemic exchanges, cardiovascular and intracerebral instability can develop. In part, the pressures generated within the dialyzer compartment by using high pump speeds can overcome continuous ripple pumps, and stop-go pumps which have been set up to control haemofiltrate and dialysate effluent rates [9].

More recently, truly volumetric pumps have been integrated into a commercially available continuous haemofiltration and/or dialysis machine, resulting in the cardiovascular and cerebral stability achieved with the spontaneously driven circuits, but with the advantages of a pumped system [10].

Dialyzer Membrane

Initial cardiovascular and cerebral instability occurring within the first hour of intermittent haemodialysis and/or haemofiltration may be due in part to dialyzer membrane bioincompatibility. In a recent study, polyacrylonitrile membranes were shown to cause fewer adverse changes in cardiovascular and cerebral stability compared to another synthetic polyamide membrane in a group of critically ill patients with FHF and established oliguric renal failure [11].

Extracorporeal Anticoagulation

Despite the reduction in clotting factors and platelets, clotting of the dialyzer can be a problem, particularly when spontaneously driven extracorporeal circuits are used.

Most centres have used traditional unfractionated heparin, low molecular weight heparin or epoprostenol as the extracorporeal anticoagulant in spontaneously driven circuits. In patients without significant risk factors for haemorrhage, standard unfractionated heparin remains the anticoagulant of choice. However, in patients with recent overt haemorrhage, epoprostenol was reported to be superior to both standard heparin [12] and low molecular weight heparin in reducing the number of bleeding episodes and so allowing extracorporeal treatment to be maintained. In addition, dialyzer life was prolonged with epoprostenol [12].

Less anticoagulation is required at the faster blood flows achieved by pumped continuous haemofiltration and/or dialysis circuits. Dialyzer life has not been significantly increased by the use of any single anticoagulant. Haemorrhage still remains a problem, and although some patients can be treated using no extracorporeal anticoagulant, they remain a small minority. In the high-risk group of patients, epoprostenol has proved to be superior to both heparin [12] and low molecular weight heparin [E.J. Will, pers. commun.] in reducing both the frequency and extent of haemorrhage.

Dialysate Anionic Base

Lactate and acetate are commonly used as the anionic buffer bases used in commercially available dialysis and haemofiltration substitution fluids. Under normal circumstances, both are converted indirectly, but on an equimolar basis, to bicarbonate – a physiological base. Studies on acetate have shown that acetate can act as a vasodilator at both the post-capillary and venous capacitance level. Therefore, in patients with acute liver failure, in whom acetate would be expected to accumulate, an additional vasodilator could have adverse effects in patients who already have marked circulatory vasodilatation.

The normal hepatic lactate clearance has been estimated at 100 mmol/h [5]. In patients with severe liver failure, due to the mismatch in tissue oxygen demand/supply, endogenous lactate production is increased and hepatic lactate uptake may be decreased. Under these circumstances, high rates of exogenous lactate administration during intermittent haemofiltration treatments produces hyperlactataemia. In severely ill patients with FHF this may be associated with an increase in arterial hydrogen ion accumulation, reduction in plasma bicarbonate [13]. Lactate may act similarly to acetate in causing further peripheral vasodilatation, and this may adversely affect cardiac output [8]. However, the amount of lactate administered during continuous modes of renal replacement therapies is much less and unlikely to result in significant hyperlactataemia in the majority of cases [5]. If hyperlactatemia does occur in association with either an increase in arterial hydrogen ion concentration or reduction in plasma bicarbonate, then bicarbonate-based replacement/dialysate should be substituted [13].

Treatment of Surges in IP during Haemofiltration and/or Dialysis

Mannitol is the standard treatment for surges in ICP in patients with an intact blood-brain barrier and normal renal function. However, in oliguric patients with acute hepatic failure, who may have cerebral oedema due both to

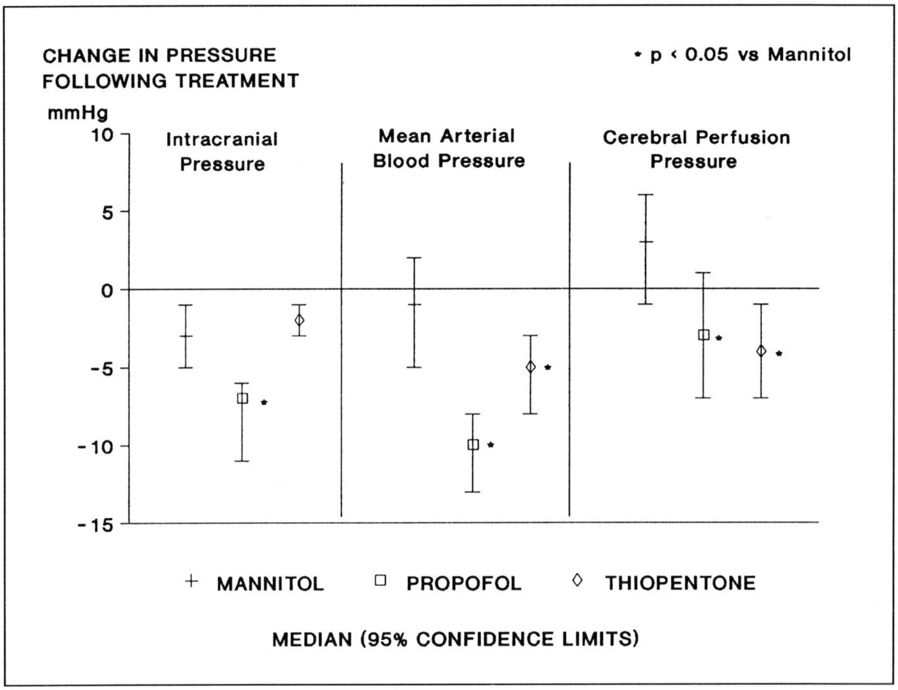

Fig. 1. Changes in IP, MAP and CPP 5 min after the administration of bolus mannitol, propofol or thiopentone.

vasogenic and neurogenic mechanisms [4], boluses of mannitol may not be so effective. Similarly, fluid removal by deliberate ultrafiltration may critically reduce the 'effective' circulating volume, resulting in reduced cardiac output and cerebral oxygen delivery, with a net effect of initially reducing ICP followed by a sustained and fatal rebound in ICP.

In a study of using bolus mannitol (100 ml of 20%), propofol (50–100 mg) or thiopentone (100–300 mg) in managing over 200 surges of ICP during isovolaemic continuous haemofiltration/haemodialysis treatments, propofol was more effective than mannitol or thiopentone in reducing ICP (fig. 1). However, the CPP was better maintained with mannitol (fig. 1). This suggests that in cases with an adequate CPP, surges of ICP should be treated with bolus propofol, whereas mannitol should be used when the CPP is 60 mm Hg or less.

References

1 O'Grady JG, Alexander GJM, Thick M, Potter D, Calne RY, Williams R: Outcome of orthotopic liver transplantation in the aetiological and clinical variants of acute liver failure. Q J Med 1988; 258:817–824.

2 Bihari DJ, Gimson AES, Williams R: Cardiovascular, pulmonary and renal complications of fulminant hepatic failure. Semin Liver Dis 1986;2:119–128.

3 Wilkinson SP, Weston MJ, Parsons V, Williams R: Dialysis in the treatment of renal failure in patients with liver disease. Clin Nephrol 1977;8:141–143.

4 Ede RJ, Gove CD, Williams R: Increased cerebral blood flow in fulminant heptic failure due to paracetamol overdose; in Soeters PB, Wilson JHP, Meijer AJ, Holm E (eds): Advances in Ammonia Metabolism and Hepatic Encephalopathy. Amsterdam, Elsevier Science, 1988, pp 567–570.

5 Davenport A: Studies in Hepatic and Renal Failure; thesis, Cambridge 1990.

6 Ravussin P, Guinard JP, Rally F, Thorin D: Effect of propofol on cerebrospinal fluid pressure and cerebral perfusion pressure in patients undergoing craniotomy. Anesthesiology 1988;43(suppl):37–41.

7 Davenport A, Will EJ, Davison AM: Early changes in intracranial pressure during haemofiltration treatment in patients with grade IV hepatic encephalopathy and acute oliguric renal failure. Nephrol Dial Transplant 1990;5:192–198.

8 Davenport A, Will EJ, Davison AM: Effect of renal replacement therapy on patients with combined acute renal and fulminant hepatic failure. Kidney Int 1993;47(suppl 41):S245–S251.

9 Davenport A, Will EJ, Davison AM: Effect of the direction of dialysate flow on the efficiency of continuous arteriovenous haemodialysis. Blood Purif 1990;8:329–336.

10 Dyson EH, Johnston P, Prabhu P, Goutcher E, Davison AM, Will EJ: Volumetric control of continuous haemodialysis in multiorgan failure. Artif Organs 1991;15:439–442.

11 Davenport A, Davison AM, Will EJ: Membrane biocompatibility: Effects on cardiovascular stability in patients on haemofiltration. Kidney Int 1993;43(suppl 41):S230–S234.

12 Davenport A, Will EJ, Davison AM: Comparison of the use of standard heparin and prostacyclin anticoagulation in spontaneous and pump-driven extracorporeal circuits in patients with combined acute renal and hepatic failure. Nephron 1994;66:431–437.

13 Davenport A, Will EJ, Davison AM: Paradoxical increase in arterial hydrogen ion concentration in patients with hepatorenal failure given lactate-based fluids. Nephrol Dial Transplant 1990;5:342–346.

Dr. A. Davenport, Southmead Hospital, Bristol BS10 5NB (UK)

Sieberth HG, Stummvoll HK, Kierdorf H (eds): Continuous Extracorporeal Treatment in Multiple Organ Dysfunction Syndrome. Contrib Nephrol. Basel, Karger, 1995, vol 116, pp 28–33

Continuous Renal Replacement Therapies: The Need for a Standard Nomenclature

Claudio Ronco[a], *Rinaldo Bellomo*[b]

[a] Department of Nephrology, St. Bortolo Hospital, Vicenza, Italy, and
[b] Intensive Care Unit, Austin Hospital, Heidelberg, Victoria, Australia

Several forms of renal replacement therapy have been used for the treatment of acute renal failure in the critically ill patient in the last decade. The evolution of basic concepts into a more complex clinical approach and applied technology, now requires a detailed analysis of the various techniques with the aim of establishing a common nomenclature.

Mechanisms of Solute Transport

Diffusion: This term defines the movement of solute with a statistical tendency to reach the same concentration of solute in the available distribution space on each side of the membrane. The practical result is a passage of molecules from the more concentrated compartment into the less concentrated compartment.

Convection: This term defines the movement of solute across a semipermeable membrane in which solute is 'carried' together with solvent across the membrane by means of a process of filtration. Filtration occurs in response to a transmembrane pressure gradient. Once ultrafiltration occurs, solutes are carried on the other side of the membrane at various rates according to their membrane rejection coefficient (σ), with σ being near 1 for albumin and near 0 for small solutes like urea. The sieving coefficient for a solute (S), correlates with the reflection coefficient according to the formula $S = 1 - \sigma$. In clinical practice, the sieving coefficient is measured from the ratio between the concentration of the solute in the ultrafiltrate and its concentration in plasma water. In convective treatments therefore, the transport depends on the volume of ultrafiltrate and the solute concentration in the ultrafiltrate. When the sieving is 1, clearance equals ultrafiltration rate.

Ultrafiltration: This is a process in which plasma water and crystalloids are separated from the whole blood across a semipermeable membrane in response to a transmembrane pressure gradient:

$$TMP = Pb - Puf - \pi,$$

where Pb is the hydrostatic pressure in the blood compartment, Puf is the hydraulic pressure in the ultrafiltrate/dialysate compartment, and π is the oncotic pressure generated by proteins.

Membranes

Different membranes are employed in renal replacement treatments. They could be divided into two categories: cellulose-based and synthetic membranes:

Cellulose-Based Membranes: These membranes (Cuprophan, Hemophan, cellulose acetate) are generally low-flux membranes having for example a permeability coefficient of $K_m < 10$ ml/h × mm Hg × m^2. They are very thin (5–15 μm of wall thickness) and they have a symmetric structure with uniform porosity. They are substantially hydrophilic.

Synthetic Membranes: These membranes (polysulfone, polyamide, polyacrylonitrile) are high-flux membranes with a $K_m > 30$ ml/h × mm Hg × m^2. Wall thickness ranges between 40 and 75 μm with an asymmetric structure composed of an inner skin layer and a surrounding sponge layer. They have large pores and are hydrophobic. These membranes have high sieving coefficients for solutes in a wide range of molecular weights and therefore they are more suitable for convective treatments. High filtration rates always take place in the presence of these membranes with their associated amount of convection. Thus, when synthetic membranes are used, extracorporeal treatment is not adequately defined by the term 'hemodialysis'. It should now be more suitably named 'hemodiafiltration' (if replacement solution is needed) or 'high-flux dialysis' if a filtration-backfiltration mechanism is present and no replacement is required.

Treatments

Hemodialysis (HD): This term defines a prevalently diffusive treatment in which blood and dialysate are circulated in countercurrent mode, and a low permeable cellulose-based membrane is employed. Ultrafiltration rate is approximately equal to the scheduled weight loss. This treatment can be performed inter-

Table 1. Schematic summary of the renal replacement therapies

Technique	Abbreviation	Diffusion	Convection	Membrane	Vascular access	Replacement fluid	Backfiltration	Tx schedule
Intermittent hemodialysis	HD	+++++	+----	Cellulosic	Fistula or VV	No	+----	3/week (4 h)
Daily hemodialysis	daily HD	+++++	+----	Cellulosic	Fistula or VV	No	+----	7/week (2–3 h)
Intermittent hemofiltration	HF	-----	+++++	Synthetic	Fistula or VV	Yes	-----	3/week (30 l)
Daily hemofiltration	daily HF	-----	+++++	Synthetic	Fistula or VV	Yes	-----	7/week (20 l)
Intermittent hemodiafiltration	HDF	+++--	+++--	Synthetic	Fistula or VV	Yes	+----	3/week (3 h, 9 l)
Daily hemodiafiltration	daily HDF	+++--	+++--	Synthetic	Fistula or VV	Yes	+----	7/week 3 h, 4–6 l)
Intermittent high-flux dialysis	HFD	++++-	++---	Synthetic	Fistula or VV	No	+++++	3/week (3–4 h)
Daily high-flux dialysis	daily HFD	++++-	++---	Synthetic	Fistula or VV	No	+++++	7/week (2–3 h)
Continuous AV hemofiltration	CAVH	-----	+++++	Synthetic	Artery-Vein	Yes	-----	Continuous
Continuous VV hemofiltration	CVVH	-----	+++++	Synthetic	Vein-Vein	Yes	-----	Continuous
Continuous A-V hemodialysis	CAVHD	+++++	+----	Cellulosic	Artery-Vein	No	-----	Continuous
Continuous V-V hemodialysis	CVVHD	+++++	+----	Cellulosic	Vein-Vein	No	-----	Continuous
Continuous AV hemodiafiltration	CAVHDF	+++--	+++--	Synthetic	Artery-Vein	Yes	-----	Continuous
Continuous VV hemodiafiltration	CVVHDF	+++--	+++--	Synthetic	Vein-Vein	Yes	-----	Continuous
Continuous AV high-flux dialysis	CAVHFD	++++-	++---	Synthetic	Artery-Vein	No	+++++	Continuous
Continuous VV high-flux dialysis	CVVHFD	++++-	++---	Synthetic	Vein-Vein	No	+++++	Continuous
Intermittent ultrafiltration	UF	-----	+++++	Synthetic	Vein-Vein	No	-----	3/week (2–3 h)
Daily ultrafiltration	daily UF	-----	+++++	Synthetic	Vein-Vein	No	-----	7/week (1–2 h)
Slow continuous ultrafiltration	SCUF	-----	+++++	Synthetic	AV or VV	No	-----	Continuous
Slow continuous UF with dialysate	SCUF-D	++---	+++--	Synthetic	AV or VV	No	-----	Continuous

mittently (4 h, 3 times/week), daily (2–4 h) or continuously, both in arteriovenous or venovenous mode (CAVHD-CVVHD) [1–4].

Hemofiltration (HF): This term defines an exclusively convective treatment with highly permeable membranes. The ultrafiltrate produced is replaced completely or in part by a sterile solution. Blood purification and fluid control is thus achieved. Any weight loss results from the differences between ultrafiltration and reinfusion rates (no dialysis fluid is used). This treatment can be performed intermittently (30 liters/session, 3 times/week) [5], daily (10–30 liters exchange) or continuously, both in arteriovenous or venovenous mode (CAVH or CVVH) (12–30 liters exchange/day) [6, 7].

Hemodiafiltration (HDF): This is a combined treatment in which diffusion and convection are combined utilizing a highly permeable membrane. Blood and dialysate are circulated as in hemodialysis but typically an ultrafiltration in excess of the scheduled weight loss is produced. To achieve fluid balance, sterile solution is reinfused in the patient at an adequate rate. This treatment can be performed intermittently (3–4 h and 9–15 liters exchange/session, 3 times/week) [8], daily (3 h and 9 liters exchange) or continuously, both in arteriovenous or venovenous mode (CAVHDF-CVVHDF) [9].

High-Flux Dialysis (HFD): This term describes a treatment that utilizes highly permeable membranes in conjunction with an ultrafiltration control system. Blood and dialysate are circulated as in hemodialysis, but due to the high permeability coefficient of the membrane, ultrafiltration would exceed the required patient weight loss. Therefore, a positive pressure is applied to the dialysate compartment to reduce the amount of ultrafiltration and to avoid the need of replacement solution. Due to the peculiar structure of hollow-fiber dialyzers, filtration takes place in the proximal part of the filter, while backfiltration occurs in the distal part. Diffusion and convection therefore are still combined because the high filtration rate occurring in the proximal part of the dialyzer is masked by a backfiltration occurring in the distal part. Replacement is avoided as it occurs inside the filter by the mechanism of backfiltration. This implies that dialysate must be sterile and pyrogen-free [10].

Ultrafiltration (UF): This term defines a treatment in which fluid removal is the main target of the therapy. Highly permeable filters are utilized and fluid is removed from the body without providing any replacement solution. UF can be performed intermittently or daily, and a maximum of 3 liters/session is generally removed in a period of 3–4 h. In clinical practice, ultrafiltration has also been used in sequence with hemodialysis to improve cardiovascular tolerance [11]. In critically ill patients, it can also be performed continuously at very low filtration rates (1–2 ml/min) with or without the addition of dialysate (SCUF or SCUF-D). Both arteriovenous and venovenous modes can be used [12].

Blood Access and Extracorporeal Circuit

Extracorporeal blood treatments require vascular access and a specific extracorporeal circuit design. Two different approaches are in use for extracorporeal treatments [13]: (1) the *arteriovenous mode:* in this system the arteriovenous pressure gradient of the patient is the driving force moving blood through the circuit, and (2) the *venovenous mode:* this type of circulation requires a roller pump in the prefilter line segment, as well as a drip chamber in the line returning the blood from the filter.

All major forms of renal replacement are summarized in table 1. It is worthy of note that all the techniques described above have long names. This has led to a veritable 'jungle' of abbreviations which confuse readers and generate misunderstandings. A table is provided which attempts to both provide a 'logical' key and adhere to currently known abbreviations.

Clinical Indications

From the above mentioned classification, one could forecast the performance of any treatment in relation to specific targets, marker molecules and fluid balance.

Ultrafiltration (UF) and slow continuous ultrafiltration (SCUF) for example are definitely indicated in patients with fluid overload or in all the oliguric states in which large amounts of intravenous infusions are required. Patients that typically benefit from these treatments are those with congestive heart failure, refractory ascites and peripheral edema unresponsive to diuretics.

All other therapies are also indicated when acute renal failure is present and the patient becomes azotemic. However, while hemodialysis (intermittent or continuous) is effective in removing small molecular weight solutes like urea, other treatments which include convective transport are generally required when other molecules must be removed. Recently, interest has grown around the possibility of removal of cytokines and mediators of inflammation in patients with septic syndromes, by means of CAVH-CVVH or CAVHDF-CVVHDF and finally continuous high-flux dialysis (CHFD). In particular, the hydrophobic structure of synthetic membranes would play a role in permitting a significant adsorption and filtration of molecules such as interleukin-1β, tumor necrosing factor-α and platelet-activating factor [14, 15]. Whether this can beneficially affect the outcome of critically ill patients is still under debate. Additionally, there is yet no definition of adequacy of renal support in the critically ill patient. As a consequence, the number of liters to be exchanged per day, the daily clearance of small and larger molecules and other treatment parameters have not yet been concretely established.

Prospective randomized trials are definitely needed to address these issues in the coming future.

Different techniques should today be considered as part of a whole range of therapeutic possibilities for the treatment of acute renal failure in the critically ill, more than opposite and opposing methods of therapy.

References

1 Lowrie EG, Sargent JA: Clinical example of pharmacokinetic and metabolic modeling: Quantitative and individualized prescription of dialysis therapy. Kidney Int 1980;18(suppl 10):11–16.

2 Buoncristiani U, Quitalini G, Cozzari M: Daily dialysis: Long-term clinical-metabolic results. Kidney Int 1988;33:(suppl 24)137–140.

3 Geronemus R, Schneider N: Continuous arteriovenous hemodialysis: A new modality for treatment of acute renal failure. Trans Am Soc Artif Intern Organs 1984;30:610–614.

4 Sigler MH, Teehan BP, Valkenburgh DV: Solute transport in continuous hemodialysis. A new treatment for acute renal failure. Kidney Int 1987;32:562–570.

5 Henderson LW: Hemofiltration for the treatment of hypertension associated with end-stage renal failure. Artif Organs 1980;4:103–107.

6 La Greca G, Fabris A, Ronco C: CAVH. Proceedings of the International Symposium on Continuous Arteriovenous Hemofiltration, Vicenza. Milano, Wichtig, 1986.

7 Ronco C, Fecondini L, Gavioli L, Conz P, Milan M, Dell'Aquila R, Bragantini L, Brendolan A, Chiaramonte S, Crepaldi C, Feriani M, La Greca G: A new blood module for continuous renal replacement therapies. Int J Artif Organs 1983;17:14–18.

8 Henderson LW, Lilley JJ, Ford CA, Stone RA: Hemodiafiltration. J Dial 1977;1:211–217.

9 Ronco C: Arteriovenous hemodiafiltration: A possible way to increase urea removal during CAVH. Int J Artif Organs 1985;8:61–62.

10 Ronco C: Continuous renal replacement therapies for the treatment of acute renal failure in intensive care patients. Clin Nephrol 1993;40:187–198.

11 Bergstrom J, Asaba H, Furst P, Oules R: Dialysis ultrafiltration and blood pressure. Proc Eur Dial Transplant Assoc 1976;13:293–299.

12 Paganini EP, Flague J, Whitman G, Nakamoto S: Amino acid balance in patients with oliguric acute renal failure undergoing slow continuous ultrafiltration. Trans Am Soc Artif Intern Organs 1982;28:615–620.

13 Bosch JP, Ronco C: Continuous arteriovenous hemofiltration; in Maher JF (ed): Replacement of renal function by dialysis. Dordrecht, Kluwer Academic, 1989, vol 15, pp 347–361.

14 Bellomo R, Tipping P, Boyce N: Tumor necrosing factor clearances during veno-venous hemodiafiltration in the critically ill. Trans Am Soc Artif Intern Organs 1991;37:322–323.

15 Ronco C, Tetta C, Camussi G, La Greca G: Platelet-activating factor removal by experimental CAVH. Crit Care Med 1995;1:99–107.

Claudio Ronco, MD, Department of Nephrology, St. Bortolo Hospital, I–36100 Vicenza (Italy)

Sieberth HG, Stummvoll HK, Kierdorf H (eds): Continuous Extracorporeal Treatment in Multiple Organ Dysfunction Syndrome. Contrib Nephrol. Basel, Karger, 1995, vol 116, pp 34–37

Comparison between Continuous Venovenous and Intermittent Hemodialysis in Acute Renal Failure

Christian Friedrichsohn, Andreas Pollak, Frank Fischer, Hans Köhler, Werner Riegel

Department of Medicine, Universitätskliniken des Saarlandes, Homburg/Saar, Germany

Several therapies may be used to treat acute renal failure but there is little information about therapies which are most appropriate in a given circumstance. Although several factors explain high acute renal failure mortality rates, there is a growing concern that the choice of renal replacement therapy may be important in affecting outcome [1, 2]. Continuous hemofiltration and hemodialysis have become popular as methods of renal replacement therapy for patients with acute renal failure since it is generally better tolerated than intermittent hemodialysis, facilitating control of the extracellular fluid volume and causing less disturbances to the cardiovascular system [3]. Barton et al. [4] have recently shown seven factors deciding about the outcome from acute renal failure which characterize multiple organ failure: the need of inotropic dosis of sympathomimetic drugs, the need of artificial ventilation, oliguric acute renal failure in comparison to nonoliguric, jaundice, older age, low serum creatinine at the initiation of hemofiltration and severe acidosis. Apache II score is a means to characterize patients suffering from multiorgan disease syndrome [5]. The ability of Apache II score to provide an objective ranging of illness severity has recently be questioned in acute renal failure [6, 7]. Therefore, we conducted the following retrospective study of 64 consecutive patients with acute renal failure on different intensive care units undergoing intermittent or continuous venovenous hemodialysis. We characterized these patients and the course of clinical outcome to get further information if there exists a qualitative difference between continuous replacement therapy and intermittent treatment.

Patients and Methods

Sixty-four patients with acute renal failure within a multiorgan disease syndrome were retrospectively analyzed. They consisted of patients who were consecutively treated by intermittent (group 1) or continuous venovenous hemodialysis (group 2). In each group, 32 patients were recorded. They were comparable for gender (group 1: 10 female, 22 male; group 2: 9 female, 23 male), age (group 1: 59 ± 14; group 2: 59 ± 14 years) and diagnosis of acute renal failure (postoperative, cardiac failure, sepsis).

Treatments were performed as follows:

Group 1: Intermittent treatment was done for 2–3 h every day. Vascular access by a single-lumen catheter was provided. Blood flow was between 200 and 300 ml/min. Heparin served as anticoagulant. Dialysate flow was 500 ml/min. Dialysis membranes consisted of cuprophane (Gambro, Lund, Sweden) or polysulfone (Fresenius, Bad Homburg, Germany).

Group 2: Venovenous treatment was performed 24 h/day. Vascular access by a double-lumen catheter was provided. Blood flow was between 60 and 100 ml/min. Heparin served as anticoagulant. Dialysate flow was 16 ml/min. Dialysis membranes consisted of AN 69 (Hospal, Lyon, France).

The parameters evaluated were: Apache II score, need of catecholamines and oxygen; analysis of blood pressure (mean arterial pressure), diuresis per 24 h, and serum creatinine values.

Data were calculated as mean ± SEM. Analyses were done at the beginning, the fifth day and at the end of the treatment period. Statistical analyses of data obtained at different time points within one group were performed by Wilcoxon's test for paired samples and analysis between the two groups at equal time points with the Mann-Whitney U test.

Results

Apache II score was 17.6 ± 0.78 in group 1 and 18.0 ± 0.63 in group 2. The score did not change significantly within the treatment time in the two groups. No differences were evaluated in either group.

Mean arterial blood pressure was significantly lower in group 2 (71 ± 2 mm Hg) versus group 1 (82 ± 3 mm Hg) ($p < 0.05$). Till the end of the observation period the significant difference between the two groups could be measured.

Diuresis was markedly lower in patients undergoing continuous hemodialysis treatment (group 2: 585 ± 137 ml/day) versus intermittent therapy (group 1: 1,181 ± 183 ml/day) ($p < 0.05$). During the treatment course, diuresis increased significantly in group 2: 868 ± 247 ml/day ($p < 0.05$). In contrast, diuresis decreased in the intermittent hemodialyzed group (group 1: 994 ± 257 ml/day; $p < 0.05$) from the beginning to day 5 in the treatment course.

Catecholamine supply to stabilize blood pressure was significantly higher in group 2 with 12 patients needing suprarenin and 9 dobutamine in comparison to group 1 with 5 patients needing suprarenin and 4 dobutamine. Norepinephrine was used in 4 patients in each group. The number of patients on catecholamines remained constant in group 2 and enhanced slightly in group 1 at the end of the treatment course.

Oxygen concentration of inspiratory air was significantly elevated at the beginning of therapy in patients undergoing continuous treatment (group: 2 0.60 ± 0.05 versus group 1: 0.44 ± 0.03; $p < 0.05$). At the end of treatment there was no significant difference between either group.

Creatinine serum values were significantly lower in group 2 (4.0 ± 0.4 mg/dl) versus group 1 (5.2 ± 0.3 mg/dl) ($p < 0.05$). The significant difference remained at the end of the treatment course.

Discussion

The data presented demonstrate that the treatment of patients with acute renal failure has changed by means of continuous treatment possibilities. Patients who were more ill could be treated. The Apache score could not characterize the severity of illness [6–8] but the need of catecholamines, the need of oxygen supply in mechanical ventilation, reduced diuresis and lower creatinine values at the beginning of kidney replacement therapy. These data fit the analysis of Barton et al. [4] which was done by a logistic regression analysis to identify important prognostic factors of outcome in 255 consecutive patients suffering from acute renal failure at an intensive care unit. They described [4] seven parameters determining outcome of patients with acute renal failure. Our patients undergoing continuous treatment (group 2) are different from those with intermittent therapy in four of these parameters at the beginning of therapy. During the treatment course, jaundice became the fifth parameter (data not shown). Age and acid-base balance which are also considered as prognostic risk parameters [4] were not different between the two groups. Therefore, we can conclude that continuous venovenous hemodialysis allows treatment of patients with a more severe degree of illness compared to intermittent hemodialysis. Furthermore, Apache II score generally correlates with survival of critically ill patients but does not seem to be a very sensitive parameter for this issue. These results are in accordance with previous reports which demonstrate tendencies to a better outcome in multiorgan failure patients when continuous hemofiltration or hemodialysis was compared to intermittent treatment [8]. By starting continuous renal replacement therapy in critically ill patients, the survival rate cannot be expected to be particularly high. Therefore, if the same outcome exists in both treatment groups, continuous treatment qualifies as a superior treatment in management of acute renal failure in critically ill patients.

References

1 Schetz M, Lauwers PM, Ferdinande P: Extracorporeal treatment of acute renal failure in the intensive care unit: A critical view. Intensive Care Med 1989;15:349–357.

2 Harris D: Acute renal replacement – Which treatment is best? Aust NZ J Med 1990;20:197–200.

3 Davenport A, Will EJ, Davidson AM: Improved cardiovascular stability during continuous modes of renal replacement therapy in critically ill patients with acute hepatic and renal failure. Crit Care Med 1993;21:328–338.

4 Barton IK, Hilton PJ, Taub NA, Warburton FG, Swan AV, Dwight J, Mason JC: Acute renal failure treated by haemofiltration: Factors affecting outcome. Q J Med 1993;86:81–90.

5 Knaus WA, Draper EA, Wagner DP, Zimmermann JE: Apache II: A severity of disease classification system. Crit Care Med 1985;13:818–829.

6 Millar BW, Boyd OF, Henderson S, Grounds RM, Benett ED: Apache II cannot be used to compare performance between intensive care units. Clin Sci 1992;82:25P.

7 Maher ER, Robinson KN, Scoble JE, Farrimond JG, Browne DR, Sweny P, Moorhead JF: Prognosis of critically ill patients with acute renal failure: Apache II score and other predictive factors. Q J Med 1989;72:857–866.

8 Kierdorf H: Continuous versus intermittent treatment: Clinical results in acute renal failure. Contrib Nephrol. Basel, Karger, 1991, vol 93, pp 1–12.

Priv.-Doz. Dr. Werner Riegel, Medizinische Universitätsklink IV,
D–66421 Homburg/Saar (Germany)

Sieberth HG, Stummvoll HK, Kierdorf H (eds): Continuous Extracorporeal Treatment in Multiple Organ Dysfunction Syndrome. Contrib Nephrol. Basel, Karger, 1995, vol 116, pp 38–47

Continuous Venovenous Hemofiltration in Acute Renal Failure: Is a Bicarbonate- or Lactate-Buffered Substitution Better?

H. Kierdorf, C. Leue, B. Heintz, J. Riehl, H. Melzer, H.G. Sieberth

Department of Internal Medicine II, Hospital of the Technical University, Aachen, Germany

Continuous venovenous hemofiltration (CVVH) is an accepted treatment modality for critically ill patients with acute renal failure (ARF) [1, 2]. Renal replacement therapy in ARF has three major aims [3]: detoxification, fluid elimination and compensation of acidosis. In commercially available fluids, lactate (30–45 mmol/l), which is converted to bicarbonate on an equimolar basis under physiological conditions, is used to combat acidosis. The lactate buffer has the advantage of greater stability over the bicarbonate buffer. However, lactate is thought to have negative effects on metabolic parameters, such as causing a decreased regeneration rate of ATP and increased protein catabolism [4].

For sufficient control of azotemia also in hypercatabolic patients, CVVH treatment requires a total of 20–30 liters hemofiltrate/day [5]. Lactate-buffered substitution solutions lead to a daily lactate load of 800–1,300 mmol. ARF patients with concomitant hepatic insufficiency or hemodynamic instability especially display an increased plasma lactate level. Reports of a reduced lactate tolerance and a tendency to develop hyperlactatemia during treatment with lactate-buffered solutions are also extent [6, 7]. Bicarbonate-buffered solutions are thought to be of potentially greater benefit in these patients [8], but are not commercially available due to the instability of solutions containing bicarbonate.

We therefore investigated the influence of a lactate-buffered and a bicarbonate-buffered substitution solution on acid-base status, electrolytes and metabolic changes in critically ill patients with ARF undergoing CVVH. The bicarbonate-buffered solution was mixed immediately before use from a buffer-free solution and bicarbonate.

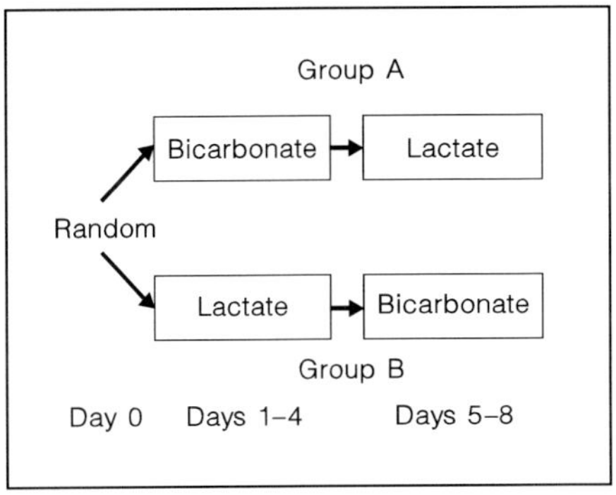

Fig. 1. Study design of the randomized crossover trial.

Subjects and Methods

Patients. Between July 1991 and June 1992, 32 consecutive patients with ARF were prospectively investigated. Patients were included if they developed acute oliguric (<400 ml urine/day) or anuric renal failure and had additional respiratory failure (mechanical ventilation, positive end-expiratory pressure >10 cm H_2O and/or a fraction of inspired oxygen (FiO_2) >0.4 [9]). In addition, organ failure for heart and liver were defined according to Goris' multiple organ failure score [9]. The clinical diagnosis of sepsis was considered according to the definition of the severe sepsis study group [10]. Patients with hepatic failure (score 1 or 2 according to Goris) were also excluded from the study as were those with pre-existing renal failure (creatinine ≥200 µmol/l), lactatemia (lactate ≥3.5 mmol/l) or known diabetes mellitus. Patients were statistically evaluated only if they finished an 8-day CVVH period.

Study Design. Patients were randomly allocated to receive bicarbonate-buffered substitution on days 1–4 and lactate-buffered substitution on days 5–8 (group A) or in reverse order (group B; fig. 1). The prospective, crossover study was approved by the hospital ethics committee in accordance with the ethical guideline of the Declaration of Helsinki 1964 as revised in Tokyo 1975.

Continuous Hemofiltration. CVVH was carried out in the postdilution mode with a BSM 22 blood pump and monitor (Hospal, France) with a constant blood flow rate of 100–140 ml/min. An AN69 polyacrylonitrile membrane (Hospal, France) was used in all cases. Filters were used as needed until the filtration rate was <600 ml/h. Vascular access was provided by internal jugular silicon twin catheters (Quinton, USA). An electronic balancing device [11] was used for the substitution fluid. Daily hemofiltration was tailored to the individual patient's need to maintain urea <20 mmol/l. The net daily fluid removal was determined by the patient's extracellular volume status.

Substitution Solutions. The final composition of the two substitution fluids is shown in table 1. Both solutions were manufactured by Schiwa, Glandorf, Germany. The bicarbonate-buffered solution was immediately mixed before use from 4.5 liters of a buffer-free solution containing sodium 109.5 mmol/l, calcium 1.81 mmol/l, magnesium 0.52 mmol/l, chloride 114 mmol/l, lactate 3.2 mmol/l, and 0.16 liter of an 8.4% bicarbonate solution (1,000 mmol/l). The lactate-buffered solution bag contained 4.5 liters.

Table 1. Composition of the substitution solutions (values in mmol/l)

	Lactate-buffered solution	Bicarbonate-buffered solution
Na^+	142	140
Cl^-	103	110
HCO_3^-	–	34.5
Lactate$^-$	44.5	3
Ca^{2+}	2.0	1.75
Mg^{2+}	0.75	0.5
Glucose	5.6	5.6

Laboratory Measurements. Arterial blood was sampled every 6 h for the measurement of base excess, pH (ABL 3; Radiometer, Copenhagen, Denmark), serum bicarbonate, sodium, potassium, phosphate, chloride, urea, creatinine (SMAC, multichannel analyzer), calcium, magnesium (atomic absorption spectrometry) and ammonia (Enzymatic Test; Du Pont, Wilmington, Del., USA). Lactate was determined using tubes containing sodium fluoride and potassium oxalate and an Enzymatic Test (Du Pont). Daily nitrogen excretion was measured in the hemofiltrate and urine by MicroKjeldahl (Auto Kjeltec 1030; Stockholm, Sweden) to determine protein catabolism.

Concomitant Treatment. All patients received an isoenergetic (25–30 nonprotein kcal/kg BW/day) and isonitrogenous (1.0–1.5 g amino acid/kg BW/day) total parenteral nutrition over the 8-day period. All patients received 40 mmol phosphate/day to avoid phosphate depletion.

Hemodynamic Effects. The mean arterial blood pressure was transduced and displayed numerically (Siemens, Göttingen, Germany). Positive inotropic substances (epinephrine, norepinephrine, dopamine or dobutamine) were allowed for each patient as needed, the mean daily dosage was measured.

Statistical Analysis. Results are expressed as mean ± SD. Statistical analysis was performed with Student's t test and the Wilcoxon rank test (paired data) and the Mann-Whitney test (unpaired data). Significance was established at $p \leq 0.05$. All p values are two-sided.

Results

Characteristics of Patients. From a total of 32 patients, 12 were eliminated from the analysis because they did not complete the 8-day CVVH period (6 in each group). Eleven patients died earlier (5 in group A) and in 1 patient CVVH was stopped because of major bleeding. Twenty patients were included in the analysis (10 in each group). The two groups were similar in terms of age, sex, etiology of ARF, severity of disease and outcome (table 2). The mean age was 61.5 ± 11 years.

Table 2. Demographic data, etiology of ARF, concomitant sepsis or heart failure and outcome

Patient No.	Age	Sex	Etiology of ARF	Sepsis or heart failure	Outcome
Group A					
1	60	f	postsurgical		alive
2	57	m	endocarditis	heart, sepsis	dead
3	69	f	pyelonephritis	sepsis	alive
4	60	m	pneumonia	sepsis	alive
5	72	f	postsurgical	heart, sepsis	alive
6	60	f	pancreatitis	sepsis	alive
7	71	f	postsurgical	heart, sepsis	dead
8	70	f	pancreatitis	heart, sepsis	dead
9	35	m	postsurgical		alive
10	44	m	myocardial infarction	heart	alive
Group B					
11	61	m	postsurgical	sepsis	alive
12	81	f	postsurgical		alive
13	64	m	postsurgical		alive
14	57	m	endocarditis	heart, sepsis	dead
15	69	m	endocarditis	heart, sepsis	dead
16	69	m	postsurgical	heart	alive
17	39	m	postsurgical	sepsis	alive
18	70	f	pneumonia	heart, sepsis	dead
19	59	m	myocardial infarction	heart	alive
20	66	f	postsurgical	sepsis	dead

Hemofiltration. There were no differences in daily hemofiltrate and net ultrafiltration rate between the two groups or between days 1–4 and 5–8 within each group. The overall daily hemofiltrate for all patients averaged 29.1 ± 8.0 liters, the net ultrafiltration rate was 2.9 ± 1.0 liters.

Hemodynamic Effects. There was no difference in mean arterial blood pressure (75.5 ± 24.0 mm Hg for all patients) or in the need of positive inotropic substances either between the two groups or within each group.

Electrolytes. There were no differences between the two groups in phosphate, potassium, chloride or sodium. This was also the case within each group, except for a small but significant ($p = 0.018$) difference in the chloride concentration in group A between part I (days 1–4) and part II (days 5–8) with a higher chloride concentration during administration of the bicarbonate buffer. A significant dif-

Table 3. Serum electrolytes (mean ± SD)

		Group A (days 1–4 bicarbonate, days 5–8 lactate)	Group B (days 1–4 lactate, days 5–8 bicarbonate)
Calcium mmol/l	Before	2.18±0.18	2.27±0.28
	Days 1–4	2.31±0.20	2.49±0.24[b]
	Days 5–8	2.58±0.26[a]	2.60±0.35
Chloride, mmol/l	Before	102±7	102±5
	Days 1–4	102±6	100±5
	Days 5–8	99±5[a]	101±5
Magnesium, mmol/l	Before	0.84±0.21	1.01±0.18
	Days 1–4	0.77±0.19	0.99±0.16[b]
	Days 5–8	0.90±0.16[a]	0.91±0.16
Phosphate, mmol/l	Before	1.12±0.74	1.61±0.76
	Days 1–4	0.84±0.45	1.11±0.48
	Days 5–8	1.10±0.30	1.14±0.39
Potassium, mmol/l	Before	4.7±0.7	4.8±0.4
	Days 1–4	4.5±0.6	4.4±0.4
	Days 5–8	4.5±0.5	4.4±0.5
Sodium, mmol/l	Before	141±7	142±4
	Days 1–4	141±4	140±4
	Days 5–8	138±3	139±6

[a] $p < 0.05$ vs. days 1–4 within group A; [b] $p < 0.05$ vs. group A.

ference could be shown on days 1–4 in calcium (p = 0.006) and magnesium concentration (p = 0.001) with the higher concentration in group B (lactate buffering), while there was no difference between the two groups on days 5–8. Within group A there was a significant increase in calcium (p = 0.007) and magnesium (p = 0.002) from days 1–4 to days 5–8 with a higher concentration during lactate substitution (table 3).

Azotemia. In all patients, urea and creatinine decreased during the 8-day period. Urea fell from 31.8 ± 16.7 mmol/l before CVVH treatment to 18.8 ± 7.1 mmol/l on day 8, and creatinine decreased from 446 ± 244 to 193 ± 91 µmol/l. There were no differences within or between the two groups (table 4).

Metabolic Parameters. No differences within or between the two groups were found for ammonia and glucose. Nitrogen excretion differed significantly between group A and B for days 1–4 (p = 0.0035), while there was no difference on days 5–8 or within each group (table 4).

Table 4. Azotemia and metabolic control (mean ± SD)

		Group A (days 1–4 bicarbonate, days 5–8 lactate)	Group B (days 1–4 lactate, days 5–8 bicarbonate)
Urea, mmol/l	Before	29.3 ± 14.1	34.3 ± 19.4
	Days 1–4	19.7 ± 8.2	24.0 ± 11.9
	Days 5–8	17.8 ± 7.1	21.3 ± 7.1
Creatinine, μmol/l	Before	397 ± 194	494 ± 288
	Days 1–4	224 ± 107	307 ± 163
	Days 5–8	173 ± 96	231 ± 68
Nitrogen excretion, g/day	Days 1–4	19.3 ± 7.6	25.5 ± 10.4[b]
	Days 5–8	20.0 ± 7.0	22.2 ± 8.0
Glucose, mmol/l	Before	11.3 ± 6.3	10.7 ± 5.1
	Days 1–4	9.7 ± 2.9	9.7 ± 2.8
	Days 5–8	9.4 ± 3.2	9.9 ± 4.4
Ammonia, μmol/l	Before	60.3 ± 40.0	44.5 ± 26.2
	Days 1–4	45.2 ± 33.8	31.6 ± 16.9
	Days 5–8	45.7 ± 33.9	34.1 ± 20.0

[b] $p < 0.05$ vs. group A.

Acid-Base Status. Renal acidosis was sufficiently controlled with no difference in pH or base excess between the groups or within the groups (table 5). On days 1–4 bicarbonate concentration was significantly higher ($p = 0.035$) in group B compared to group A, while the difference was not significant on days 5–8 ($p = 0.131$). Within group A bicarbonate was significantly higher on days 5–8 ($p = 0.043$), while within group B bicarbonate was significantly ($p = 0.018$) higher on days 1–4 (table 5). The lactate concentration was significantly reduced during bicarbonate treatment. On days 1–4 the lactate level was increased in group B ($p = 0.002$), on days 5–8 in group A ($p = 0.0018$). Within group A there was a significant difference between days 1–4 with bicarbonate (days 1–4) and lactate substitution (days 5–8; $p = 0.0018$). A difference within group B failed to reach statistical significance ($p = 0.052$; table 5). Comparing lactate values before CVVH treatment in both groups, there was a slight decrease in lactate, the decrease being most marked during bicarbonate substitution.

Table 5. Acid-base status (mean ± SD)

		Group A (days 1–4 bicarbonate, days 5–8 lactate)	Group B (days 1–4 lactate, days 5–8 bicarbonate)
pH	Before	7.38 ± 0.06	7.43 ± 0.07
	Days 1–4	7.43 ± 0.03	7.43 ± 0.03
	Days 5–8	7.43 ± 0.05	7.42 ± 0.07
Base excess	Before	–2.80 ± 3.6	0.01 ± 3.9
	Days 1–4	–0.59 ± 2.7	0.08 ± 2.2
	Days 5–8	–0.13 ± 3.1	–0.89 ± 2.3
Bicarbonate, mmol/l	Before	19.9 ± 4.4	23.5 ± 3.9
	Days 1–4	22.5 ± 2.5	23.8 ± 2.5[b]
	Days 5–8	23.2 ± 3.0[a]	22.5 ± 2.3[c]
Lactate, mmol/l	Before	2.54 ± 2.49	2.49 ± 1.04
	Days 1–4	0.76 ± 0.86	1.59 ± 1.17[b]
	Days 5–8	2.05 ± 1.21[a]	1.25 ± 0.83[b]

[a] $p < 0.05$ vs. days 1–4 within group A; [b] $p < 0.05$ vs. group A; [c] $p < 0.05$ vs. days 1–4 within group B.

Discussion

All critically ill patients with ARF were effectively treated by CVVH using either bicarbonate-buffered or lactate-buffered solutions. Within 48 h, uremia was well controlled in both groups and hyperhydration was well balanced with a daily hemofiltrate of about 29 liters, in accordance with the data of other groups using lactate-buffered substitution solutions [1, 2]. Until the present the use of bicarbonate-buffered solutions has only been analyzed in a few ARF patients undergoing intermittent hemofiltration [12] or CVVH [13]. Both studies investigated the influence of bicarbonate-buffered solutions on acid-base status without comparison to a lactate-buffered solution. In accordance with these studies, electrolytes were balanced in both groups examined in our study. The small differences in chloride, calcium and magnesium concentration between the two groups were related to the slightly different composition of the two solutions.

The composition of the bicarbonate buffer, readily mixed before use, is determined by the chemical properties of the compounds. The solution must be stable for a 24-hour period without precipitation of calcium carbonate or magnesium carbonate [13]. For this reason it is not possible to employ a nonbuffered solution. The bicarbonate buffer solution contains 3 mmol/l lactate to prevent precipitation. The magnesium and calcium concentration is reduced compared to the lactate-buffered solution. To adjust ionic strength the chloride concentration must be increased. Possible precipitation does not allow a higher phosphate concentration in one of the solutions, so phosphate must separately be substituted to avoid phosphate depletion. The readily mixed bicarbonate-buffered solution was prepared in bags of special plastic sheeting to prevent evaporation of carbon dioxide [13]. None of the patients suffered any side effects from any of the solutions.

Lactate infusions have been reported to induce a higher degree of protein catabolism than bicarbonate [4]. In our study, nitrogen excretion was significantly increased in patients receiving lactate buffer on days 1–4, while in the second part of the study on days 5–8 there was no significant difference. This is probably due to the higher protein catabolic rate in critically ill patients on the first 3–5 days [14], so that on days 5–8 the buffer may not be as important as in the first 4 days of treatment. One of the reasons for increased protein catabolism accompanying hyperlactatemia is an increased lactate metabolism in skeletal muscles which has been described in patients with increased lactate levels and in patients with decreased lactate metabolism in the liver [15].

Protein catabolism is also due to the blockade of the conversion of pyruvate to oxaloacetic acid, a mechanism which prevents intracellular gluconeogenesis [16]. There were no differences in carbohydrate metabolism between the two groups. Both solutions contained the same amount of glucose (1 mmol/l).

Especially acetate but also lactate buffer have been shown to exert a negative influence on mean arterial blood pressure and cardiac function [12, 16, 17]. During hyperlactatemia (lactate $\geq$ 3.5–4 mmol/l) a reduced myocardial performance [15] due to an increased intracellular lactate concentration, leading to reduced cellular ATP production, has been reported. A correlation of mean arterial blood pressure to the degree of lactate intolerance has also been described. In our study there was no negative influence of lactate buffer, the hemodynamic status of the patients was comparable using either the bicarbonate- or lactate-buffered solutions. No differences in mean arterial blood pressure or in the need of positive inotropic substances between or within the groups was apparent.

This blood pressure stability is probably due to the fact that in our patients, despite a significantly higher lactate level during the substitution with lactate buffer, lactate levels never exceeded the normal limits. Davenport et al. [12] have reported hyperlactatemia during lactate-buffered substitution in critically ill patients with ARF comparable to ours. Davenport et al. administered intermittent hemofiltration with a lactate load of 190–210 mmol/h, whereas our patients received a maximum lactate load of 90 mmol/h, even when hemofiltration increased to 2 liters/h. The slower ultrafiltration in CVVH compared to intermittent hemofiltration may be the reason for the better lactate tolerance in critically ill patients, despite the same total quantity of substituted lactate (1,000–

1,200 mmol/day). In contrast to the study of Davenport et al. [12], where a tendency to metabolic acidosis in the ARF group was described, renal acidosis was sufficiently controlled in our patients, with no differences between bicarbonate- or lactate-buffered solutions. Lactate could be shown to be an effective bicarbonate generating base in patients with ARF or chronic renal failure [12, 18]. Under stable clinical conditions, lactate is metabolized to bicarbonate on an equimolar basis. Critically ill patients with ARF, especially with concomitant sepsis or circulatory shock, have been reported to display a reduced lactate tolerance [7]. In our study, which excluded patients with liver disease, the only sign of a possible lactate intolerance was the significantly higher lactate levels during lactate buffering. The bicarbonate levels were also significantly higher during lactate buffering. This is probably due to a lower quantity of buffer substances in the bicarbonate-buffered solution (34.5 mmol/l bicarbonate and 3 mmol/l lactate) compared to the lactate-buffered solution (44.5 mmol/l lactate) but may also be explained by the physiological metabolism of lactate to bicarbonate, as lactate was also elevated in these patients.

In conclusion, CVVH treatment in critically ill patients with ARF using either bicarbonate-buffered or lactate-buffered solutions sufficiently controlled uremia and renal acidosis. Lactate rose during the use of lactate buffer and there was increased protein catabolism using the lactate buffer during the first 4 days. But the lactate was well metabolized even by ARF patients with concomitant sepsis or circulatory shock. Bicarbonate-buffered solutions may be useful in patients with established hyperlactatemia or in critically ill patients with impaired lactate metabolism, especially in cases with increased protein catabolism.

References

1 Canaud G, Carred LJ, Christol JP, Aubas S, Beraud JJ, Mion C: Pump-assisted continuous venovenous haemofiltration for treating acute uremia. Kidney Int 1988;33:S154–S156.

2 Wendon J, Smithies M, Sheppard M, Bullen K, Tinker J, Bihari D: Continuous high volume venovenous haemofiltration in acute renal failure. Intensive Care Med 1989;15:358–363.

3 Ronco C, Burckardi H: Management of acute renal failure in the critically ill patient; in Pinsky MR, Dhainant JFA (eds): Pathophysiologic Foundations of Critical Care. Baltimore, Williams & Wilkins, 1993, pp 630–677.

4 Veech L: The untoward effects of the anions of dialysis fluid. Kidney Int 1988;34:587–597.

5 Storck M, Hartl WH, Zimmerer E, Inthorn D: Comparison of pump-driven and spontaneous continuous haemofiltration in postoperative acute renal failure. Lancet 1991;337:452–455.

6 Cohen RD, Iles RA: Lactic acidosis. Clin Endocrinol Metab 1980;9:513–527.

7 Giblert GM, Haupt MT, Calson RW: Lactate predicts the relationship between oxygen transport and oxygen consumption in patients with circulatory shock. Crit Care Med 1984;12:299.

8 Golper TA: Continuous arteriovenous haemofiltration in acute renal failure. Am J Kidney Dis 1985;6:373–386.

9 Goris RJA, te Boekhorst TPA, Nuytinck JKS, Gimbrère JSF: Multiple organ failure. Arch Surg 1985;120:1109–1115.

10 Bone RC, Fisher CJ Jr, Clemmer TP, Slotman GJ, Metz CA, Balk RA: A controlled trial of high-dose methylprednisolone in the treatment of severe sepsis and septic shock. N Engl J Med 1987; 317:653–658.
11 Schultheis R, Brings W, Glöckner WM, Sieberth HG: Device for controlled cyclic substitution during spontaneous filtration; in Sieberth HG, Mann H (eds): Continuous Arteriovenous Hemofiltration. Basel, Karger, 1985, pp 64–66.
12 Davenport A, Will E, Davison AM: The effect of lactate-buffered solutions on the acid-base status of patients with renal failure. Nephrol Dial Transplant 1989;4:800–804.
13 Olbricht CJ, Huxmann-Nägeli D, Bischoff H: Bikarbonat- statt laktatgepufferter Substitutionslösung zur kontinuierlichen Hämofiltration im Intensivbereich. Anästh Intensivther Notfallmed 1990;25:164–167.
14 Bartlett RH, Bosch J, Geronemus R, Paganini E, Ronco C, Swartz R: Continuous arteriovenous hemofiltration for acute renal failure. Trans Am Soc Artif Intern Organs 1988;34:67–77.
15 Kreisberg RA: Mechanism of lactic acidosis. Lancet 1974;ii:960.
16 Saman S, Opie LH: Mechanism of reduction of action potential duration of ventricular myocardium by exogenous lactate. J Mol Cell Cardiol 1984;10:659–662.
17 Mansell MA, Morgan SH, Moore L, Kong CH, Laker MF, Wing AJ: Cardiovascular and acid-base effects of acetate and bicarbonate haemodialysis. Nephrol Dial Transplant 1987;1:229–232.
18 Hayashi JA, Dauirdas JT: Use of lactate as a base in hemodialysis. Am J Nephrol 1987;7:434–439.

Dr. H. Kierdorf, Department of Internal Medicine II, Hospital of the Technical University, Pauwelsstrasse 30, D–52074 Aachen (Germany)

Sieberth HG, Stummvoll HK, Kierdorf H (eds): Continuous Extracorporeal Treatment in Multiple Organ Dysfunction Syndrome. Contrib Nephrol. Basel, Karger, 1995, vol 116, pp 48–52

Removal of Methotrexate by Continuous Venovenous Hemodiafiltration

P. Jambou[a], J. Levraut[a], C. Favier[a], C. Ichai[a], G. Milano[b], D. Grimaud[a]

[a] Department of Anesthesiology and Intensive Care, University Hospital, and
[b] Department of Oncopharmacology, Centre A.-Lacassagne, Nice, France

High-dose methotrexate (MTX) regimens are widely used in the treatment of malignancy [1]. However, adequate hydration, urinary alkalinization and folinic-acid rescue adapted to residual blood MTX concentrations are necessary to prevent severe toxicity (myelosuppression, mucositis, impairment of renal function) [2]. In spite of these measures, significant toxicity and occasional fatalities (up to 6%) can occur unexpectedly [3]. MTX cytotoxicity, related to both concentration and duration of exposure, seems directly related to the maintenance of a plasma concentration [C] at T 48 h $> 2 \cdot 10^{-7}$ mol$\cdot$l^{-1} [1]. So, when potentially toxic MTX blood concentrations are encountered, prompt attempts to remove the drug from the circulation seem justified to decrease the risk of myelosuppression. Nevertheless, the respective merits of different forms of extrarenal purification (peritoneal dialysis [4], hemodialysis [5], hemodiafiltration [6], charcoal hemoperfusion [7–9], exchange transfusion [10] as means of removing MTX) remain much debated. We would like to relate 2 cases of using continuous venovenous hemodiafiltration (CVVHD) as a specific treatment of overexposure to MTX.

Cases' Medical History

Patient 1. A 12-year-old boy was treated in 1991 for retarded acute lymphoblastic leukemia relapse. The management protocol included three successive sequences (R1, R2, R3 blocks) repeated 3 times before maintenance therapy. Only the R1 and R2 blocks include a 5-g/m^2 dose of MTX on infusion day 1 (D1). The first sequence was performed with no undue toxicity and the MTX pharmacokinetic profile was in the normal range according to the age of the child and the MTX dose. When the second R1 block started, an abnormal MTX pharmacokinetic profile appeared ([$C_{T120\,h}$] = $1.5 \cdot 10^{-6}$ mol$\cdot$l^{-1}, $T_{1/2}$ = 203 h). At this time, the patient did not show signs of severe renal impairment (diuresis 1,700 ml/24 h, serum creatinine 80.5 μmol/l, creatinine clearance 50 ml/min) and no myelosuppression. However, because of slow plasma clearance of MTX and development of mucositis, a salvage protocol including intensified folinic-acid rescue and artificial drug removal with CVVHD was started on day 6 (D6) after therapy.

Patient 2. A 9-year-old boy was treated in 1992 for malignant lymphoma relapse with a 5-g/m^2 dose of MTX on infusion D1. Despite adequate hydration, the patient developed, on the second day of treatment, a nonoliguric acute renal failure (diuresis 1,200 ml/24 h, serum creatinine 348 μmol/l, creatinine clearance <6 ml/min). Because of moderate myelosuppression (leukocytes 2,300/mm^3, platelets 70,000/mm^3) and the presence of toxic concentration of MTX ($[C_{T\ 50\ h}] = 9.3 \cdot 10^{-5}$ mol·l^{-1}) associated with acute renal failure and slow plasma clearance of MTX ($T_{1/2}$ = 69 h), the patient underwent CVVHD on day 3 (D3) after therapy.

Materials and Methods

After femoral cannulation of an 8-French double-lumen hemodialysis catheter, CVVHD was performed in both patients via a venovenous circuit using a double-head pump monitoring module (BSM 22/VPM®; Hospal, Basel, Switzerland), and a 0.43-m^2 flat plate polyacrylonitrile hemofilter (AN-69S®; Hospal, France) changed every 24 h. Blood line thrombosis was prevented using a continuous infusion of standard heparin (10–15 IU/kg/h) and blood flow was kept at a constant rate of 100 ml/min. Bicarbonate dialysate (Biosedra®; France) was delivered by a volumetric countercurrent pump at a rate of 1 liter/h and replacement fluid was administered in a postdilution mode according to a central venous pressure monitoring. The ultrafiltrate was collected in a volumetric bag and measured hourly.

Results

Patient 1. MTX plasma concentrations are depicted in figure 1. The ultrafiltrate rate ranged between 700 and 1,000 ml·h^{-1}. Ultrafiltrate and urine were not collected for measurements of MTX concentrations. However, on a semi-log curve, one can see a sudden drop in the MTX plasma concentration and in half-life of MTX clearance (from 203 to 26 h) as soon as CVVHD was started, when there was no significant modification of the renal function (diuresis 1,700–2,000 ml/24 h and creatinine renal clearance 50–78 ml·min^{-1} over the CVVHD period). 92 h later, MTX plasma concentration was $1.6 \cdot 10^{-7}$ mol·l^{-1} and CVVHD was stopped. Although there was a temporary and limited rebound of MTX concentration ($2 \cdot 10^{-7}$ mol·l^{-1}) after cessation of the procedure, the patient regained normal range value of MTX concentration at D15, and was discharged from the hospital at D20.

Patient 2. MTX plasma concentrations are depicted in figure 2. For this patient, blood samples were collected from both venous and arterial blood lines,

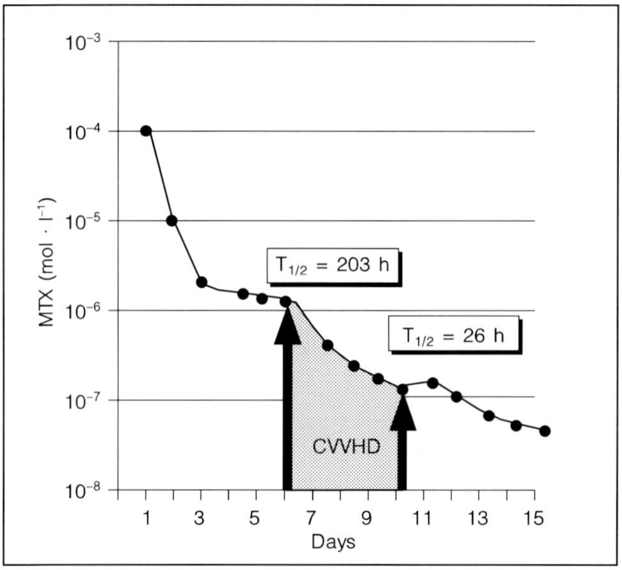

Fig. 1. MTX concentrations versus time from D1 to D15 for patient 1.

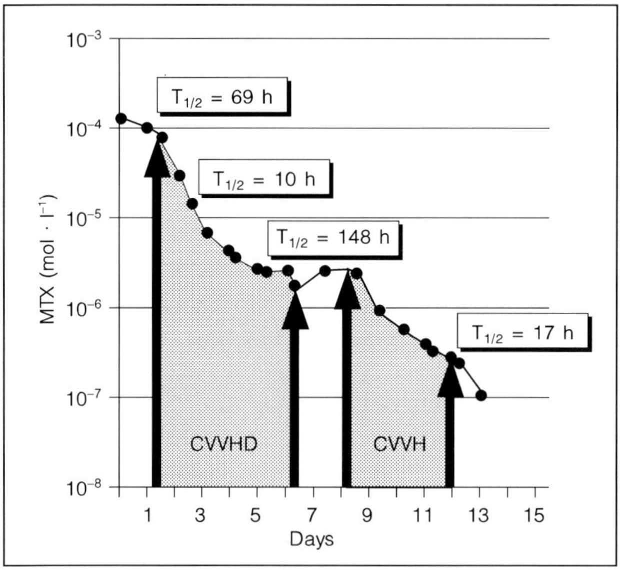

Fig. 2. MTX concentrations versus time from D1 to D15 for patient 2.

Table 1. Evolution of MTX renal and filter clearance for patient 2

	Days				
	4	5	6	10	13
MTX renal clearance, $ml \cdot min^{-1}$	23.81	12.94	37.05	53.64	182.3
MTX filter clearance, $ml \cdot min^{-1}$	27.14	27.27	23.2	21	25
Ratio filter/renal clearance, %	≈53	≈68	≈38	≈28	≈12

and samples of urine and dialysate/ultrafiltrate were taken in order to calculate clearances (table 1). The ultrafiltrate rate ranged from 900 to 1,000 $ml \cdot h^{-1}$ over the 5-day period of CVVHD. MTX plasma concentration and half-life of MTX clearance decreased (from 69 to 10 h) as soon as CVVHD was started. But, 3 days later, a steady state appeared and CVVHD seemed to be ineffective, as shown by a plateau MTX plasma concentration ($1.6 \cdot 10^{-6}$ $mol \cdot l^{-1}$) and a total half-life of MTX clearance at 148 h. So, CVVHD was stopped at D6. This resulted in an increase of MTX concentrations at $2.8 \cdot 10^{-6}$ $mol \cdot l^{-1}$ in the following 2 days, and prompted the reintroduction of continuous venovenous hemofiltration (CVVH) procedure. When CVVH was started (blood flow at 100 $ml \cdot min^{-1}$, ultrafiltrate rate ranged from 1,100 to 1,900 $ml \cdot h^{-1}$), MTX concentrations decreased and half-life of MTX clearance reduced to 17 h. After a 4-day period of CVVH, the patient regained acceptable levels of MTX concentration by D15 and normal renal function by 10 days.

Discussion

Some reports have suggested that conventional hemodialysis alone is useful only for high MTX concentrations [8] and that the most efficient method of MTX removal was a combination of hemodialysis and filtration-adsorption on charcoal [7–9]. However, this procedure is apparently limited by a nonnegligible reduction in circulating platelets and by charcoal system saturability during blood passage. Exchange transfusion is also effective in removing MTX from circulation [10], but this approach is again limited by the high blood consumption and the risk of viral transmission. Hemodiafiltration efficiently remove MTX both at high ($1.6 \cdot 10^{-6}$ $mol \cdot l^{-1}$) and at relatively low concentrations ($2 \cdot 10^{-7}$ $mol \cdot l^{-1}$) [6]. However, after all these methods, more or less marked rebound in MTX blood concentrations occur, and may be attributed to the release of MTX from storage tissues and/or erythrocytes. So, continuous and longer removal periods appear to be necessary to avoid drug rebounds. CVVHD would be particularly good in reaching this goal.

In our 2 patients, concentrations and half-lives dropped as soon as MTX removal was started, thus reducing the length of exposures to toxic substance. No rebound effect was observed when purification was stopped at [C] $<2 \cdot 10^{-7}$ $mol \cdot l^{-1}$. No secondary effect linked to this technique was noted. So, given its effectiveness and limited risks, CVVHD can be proposed as a therapeutic measure for overexposure to MTX.

References

1 Jolivet J, Cowan KH, Curt GA, Clendeninn NJ, Chabner BA: The pharmacology and clinical use of methotrexate. N Engl J Med 1983;309:1094–1104.

2 Shen DD, Azarnoff DL: Clinical pharmacokinetics of methotrexate. Clin Pharmacokinet 1978;3: 1–13.

3 Von Hoff DD, Penta JS, Helman LJ, Slavik M: Incidence of drug-related deaths secondary to high-dose methotrexate and citrovorum factor administration. Cancer Treat Rep 1977;61:745–749.

4 Ahmad S, Shen F, Bleyer AW: Methotrexate-induced renal failure and ineffectiveness of peritoneal dialysis. Arch Intern Med 1978;138:1146–1147.

5 Hande KR, Balow JE, Drake JC, Rosenberg SA, Chabner BA: Methotrexate and hemodialysis. Ann Intern Med 1977;87:495–496.

6 Montagne N, Milano G, Caldani C, Bracco J, Ayela P, Cassuto E, Thyss A, Schneider M: Removal of methotrexate by hemodiafiltration. Cancer Chemother Pharmacol 1989;24:400–401.

7 Bouffet E, Frappaz D, Laville M, Finaz J, Pinkerton CR, Philip T, Brunat-Mentigny M: Charcoal haemoperfusion and methotrexate toxicity. Lancet 1986;i:1497.

8 Djerassi I, Ciesielka W, Kim JS: Removal of methotrexate by filtration-adsorption using charcoal filters or by hemodialysis. Cancer Rep 1977;61:751–752.

9 Relling MV, Stapleton B, Ochs J, Jones DP, Meyer W, Wainer IW, Crom W, McKay CP, Evans WE: Removal of methotrexate, leucovorin, and their metabolites by combined hemodialysis and hemoperfusion. Cancer 1988;62:884–888.

10 Frappaz D, Bouffet E, Biron P, Latour JF, Kassir A, Philip T, Brunat-Mentigny M: Intoxication au méthotrexate: intérêt de l'exanguino-transfusion. Pédiatrie 1987;42:257–260.

P. Jambou, MD, Department of Anesthesiology and Intensive Care, University Hospital, F–06200 Nice (France)

Sieberth HG, Stummvoll HK, Kierdorf H (eds): Continuous Extracorporeal Treatment in Multiple Organ Dysfunction Syndrome. Contrib Nephrol. Basel, Karger, 1995, vol 116, pp 53–55

Hemofiltration in a Nephrotic Syndrome of Unknown Origin

Johannes Reiger

Medizinische Abteilung, LKH, Stolzalpe, Austria

Nephrotic syndrome results from alterations in the permselective characteristics of the glomerular basal membrane (GBM) that allows an increased passage of proteins into the urine. It consists of the constellation of heavy proteinuria, (>3.5 g/24 h), hypoalbuminemia, hyperlipoproteinemia, increased serum concentrations of high-molecular-weight proteins and edema formation [1].

The mean value for proteinuria in a number of nephrotic syndrome studies is about 8.0 g/24 h. Proteinuria in excess of 3.5 g/24 h is predictive of any of several serious renal diseases (table 1) [2]. Complications in nephrotic syndrome are mainly thrombosis, infections caused by a decrease of immunoglobulins, especially IgG and IgA, hypocalcemia, a microcytic hypochromic anemia and increased cardiovascular risk caused by the hyperlipidemia, especially cholesterol [2]. Pharmacological means to reduce proteinuria are done by ACE inhibitors, cyclooxygenase inhibitors, corticosteroids and diuretics, mainly furosemide [3, 4]

Case Report

A 69-year-old woman noticed an increase of body weight and edema of the lower extremities. Ambulant treatment with suspicion of a heart failure due to digoxin and diuretics did not show any success. Several days before admission to the hospital there had been eyelid edema and increased dyspnea. After admission the laboratory findings showed elevated urea and creatinine, a heavy urinary albumin excretion, heavy elevation of blood sedimentation rate and hyperlipidemia (table 2). It was therefore possible to confirm the diagnosis of a nephrotic syndrome.

In the chest X-ray signs of a fluid lung were found, sonographic findings were small pleural effusions on both sides and ascites. Both kidneys seemed to be without pathological findings. Despite dopamine (3 μg/kg BW/min), parenterally given nitroglycerin because of a

Table 1. Renal diseases

Diseases confined to the kidney	Systemic diseases
Minimal change nephrotic syndrome	Diabetes mellitus
Membranous nephropathy	Schönlein-Henoch purpura
Focal segmental glomerulosclerosis	Systemic lupus erythematosus
Mesangial proliferative glomerulonephritis	Amyloidosis
Mesangiocapillary glomerulonephritis	Goodpasture's syndrome
Acute poststreptococcal glomerulonephritis	*Hereditary disorders*
	Congenital nephrotic syndrome
	Hereditary nephritis (Alport's syndrome)
	Partial lipodystrophy

Table 2. Laboratory findings

Urea	117 mg/dl
Creatinine	1.52 mg/dl
Urinary albumin excretion	17.2 g/24 h
Serum albumin	14 g/l
Osmolarity	300 mosm/kg
BSR	74/108 mm
CRP	3.4 mg/dl
Fibrinogen	769 mg/dl
Cholesterol	473 mg/dl
Central venous pressure	9 cm H_2O

heart failure (4–6 mg/h), high doses of furosemide (up to 1,000 mg/day) and supplementation of albumin and protein solutions, it was not possible to remove the edema and ascites. It was decided to use hemofiltration as a support to remove the edema. In two sessions lasting approximately 6 h, 5,000 ml were removed each time using a Gambro HFM-10 pump and an HF 77 filter. After the second hemofiltration the demand for furosemide was lowered first to 120 mg/day and after several days to 80 mg/day administered parenterally.

Sonographic controls showed no ascites or pleural effusions. Altogether, the patient was admitted for 34 days at our hospital, 19 of which in the ICU. Loss of weight had been 25 kg, urinary albumin excretion had dropped to 5.6 g/24 h, serum albumin increased to 22 g/l and urea and creatinine were in the normal range. At follow-up 4 weeks later, there was no edema and no signs of a heart failure, but again an increase in urinary albumin excretion was noticed (8.5 g/24 h). Therefore, the oral prednisolone dose was increased to 50 mg/day (before, 10 mg/day had been prescribed). The final control visit 20 weeks after the first admission to the hospital showed, with the exception of cholesterol, laboratory findings in the normal range (table 3).

Table 3. Laboratory findings after 20 weeks

BSR	9/30 mm
CRP	1.3 mg/dl
Urea	40 mg/dl
Creatinine	0.9 mg/dl
Urinary albumin excretion	800 mg/24 h
Albumin	33 mg/dl
Cholesterol	311 mg/dl

Current treatment includes captopril 2 × 25 mg/day, furosemide 1 × 40 mg/day and prednisolone 10 mg/day, which are then slowly tapered off. Diagnostically, neither immunological nor infectious reasons were found for the nephrotic syndrome. A biopsy of the kidneys was not done.

Conclusion

In patients with fluid overload, hemofiltration is a better alternative to medical treatment with high doses of diuretics, especially CVVH. After the removal of only a few liters, the demand of diuretic drugs is lowered, so that medical treatment can show success. In the present case a continuous hemofiltration was not possible because of too few nurses at our ICU; we therefore decided to use intermittent hemofiltration as an initial support.

References

1 Early LE, Havel RJ, Hopper J, Grans H: Nephrotic syndrome, Calif Med 1971;115:23.
2 Kaysen GA: Proteinuria and the nephrotic syndrome; in Schrier RW (ed): Renal and Electrolyte Disorders. Boston, Little, Brown, 1992.
3 Freeman RM: Genitourinary Tract; in Schroeder StA, Tierney LM, McPhee StJ, Papadakis MA, Krupp MA (eds): Current Medical Diagnosis and Treatment. Los Altos, Lange, 1992.
4 Girndt J: Therapeutische Verfahren; in Girndt J (ed): Nieren und Hochdruckkrankheiten. Stuttgart, Schattauer, 1990.

Dr. Johannes Reiger, 2. Medizinische Abteilung, Intensivstation LKH Klagenfurt, A–9020 Klagenfurt (Austria)

Sieberth HG, Stummvoll HK, Kierdorf H (eds): Continuous Extracorporeal Treatment in Multiple Organ Dysfunction Syndrome. Contrib Nephrol. Basel, Karger, 1995, vol 116, pp 56–61

Influence of Venovenous Hemofiltration on Posttraumatic Inflammation and Hemodynamics

Werner Riegel[a], *Thomas Ziegenfuss*[b], *Stefan Rose*[c], *Michael Bauer*[b], *Ingo Marzi*[c]

Departments of
[a] Medicine,
[b] Anesthesiology and
[c] Trauma Surgery, Universitätskliniken des Saarlandes, Homburg/Saar, Germany

Secondary organ dysfunction after severe trauma or sepsis is mainly caused by a systemic inflammatory response syndrome (SIRS). SIRS is the pathophysiological reason for the development of multiorgan dysfunction syndrome (MODS) [1]. Sepsis and trauma are associated with a release of numerous mediators like endotoxins, cytokines and platelet-activating factor [2]. Most of the cytokines involved (TNF, IL-1, IL-6 and IL-8) have molecular weights lower than the cut-off of membranes used for continuous hemofiltration, e.g. IL-8 (8–10 kD), TNF and IL-1 (about 17 kD) and IL-6 (21–28 kD) [3]. Removal of cytokines by hemofiltration was recently shown by different authors and supports the preferential implementation of existing hemofiltration therapies in patients with renal failure associated with sepsis [4–7].

SIRS, for example induced by trauma, will result in different lesions subsequently. Within the first few hours after trauma, endothelial cells, polymorphonuclear cells and macrophages will be activated [8]. Disturbances of microcirculation also in primarily intact organs [9], endothelial lesions by proteases and oxygen radicals via peroxidation of membrane lipids [10] are some of the pathological pathways leading to MODS accompanied by stimulation of hepatic acute phase protein synthesis [11].

SIRS will also lead to hypotension and tachycardia which are results from a hyperdynamic circulation (enhanced cardiac output) and decreased systemic cardiovascular resistance [12]. Oxygen delivery (D_{O_2}) is enhanced in most situations but oxygen cannot be used due to disturbances in microcirculation and oxygen tissue extraction [13, 14].

Positive results of continuous hemofiltration in the course of septic shock [15–17] and the disturbed hemodynamic situation [18] have been demonstrated. Since the pathophysiology of SIRS after trauma and sepsis have some common aspects, the following study was undertaken to evaluate the role of continuous venovenous hemofiltration (CVVH) on disturbances associated with SIRS after polytrauma.

Methods

Twenty-four patients were prospectively investigated. The study was allowed by the local ethics committee. Patients with polytrauma were included into the study when the injury severity score (ISS) was ≥ 27 [19]. Patients with kidney failure, intracranial bleeding and contraindications against low-dose heparinization were excluded. The informed consent of the patients or of the relatives was obtained. The patients were randomly assigned either to the control group with standard intensive care therapy (12 patients) or the hemofiltration group with additional CVVH.

Patients included into the hemofiltration group received a double-lumen catheter. CVVH was performed with a filter containing AN 69 membrane (Hospal, Lyon, France) which was changed daily. Blood flow of 60–100 ml/min was used and the filtrate of about 500 ml/h was substituted by an electrolyte solution (SH04; Schiwa, Germany). Heparinization was the same in both groups in dosages used for thrombosis prophylaxis, i.e. 500 U/h. CVVH was started within 24 h after trauma and continued for 5 days. The investigation period was up to 2 weeks. Statistical analysis was performed by the Mann-Whitney U test and significance was accepted with a $p < 0.05$.

Results

CVVH was performed without any complications during clinical application. All patients of both groups survived the 14-day investigation period.

Mean arterial blood pressure, mean pulmonary artery pressure, central venous pressure and pulmonary capillary wedge pressure did not differ between both groups. In contrast, in the group undergoing CVVH, the substantially increased cardiac output was significantly attenuated from day 2 to day 5. e.g. 8.2 ± 0.5 (CVVH) vs. 10.3 ± 0.4 (Control) l/min, $p < 0.01$, at day 3.

Systemic vascular resistance was decreased substantially in both groups compared to normal values but less by CVVH, e.g. 839 ± 53 (CVVH) vs. 569 ± 32 (control) dyn·s/cm, $p < 0.01$, at day 3.

D_{O_2} was significantly lower from day 1 to day 5 in the CVVH group. However, oxygen consumption remained constant. Hence, oxygen extraction rate was increased in the CVVH group.

Biochemical markers of polymorphonuclear leukocytes like elastase and myeloperoxidase were decreased initially (elastase) or were decreased by CVVH treatment during the whole period of observation (myeloperoxidase, data not shown).

C-reactive protein was significantly enhanced from day 2 to day 5 in the CVVH-treated group, e.g. 179 ± 13 (CVVH) vs. 110 ± 13 (control) mg/l, $p < 0.05$, at day 2.

No change was observed for IL-6.

TNF-α was elevated (>30 pg/ml) in 8 of 12 patients in the control group but only in 3 of 12 patients of the CVVH group.

Malondialdehyde, a stable product of plasma lipid peroxidation, showed significantly lower levels from day 1 to day 11, e.g. 1.36 ± 0.24 (CVVH) vs. 2.04 ± 0.49 (Control) mg/l, $p < 0.05$, at day 5.

Discussion

The study clearly demonstrates that CVVH influenced the disturbances associated with SIRS after polytrauma. All patients showed a hyperdynamic circulatory situation, i.e. lowered peripheral resistance and elevated cardiac output, as it is typical for hyperdynamic SIRS. Patients undergoing CVVH had higher peripheral resistance and lower cardiac output compared to the control group. The importance of peripheral resistance was shown by Groeneveld et al. [20]. They demonstrated that peripheral resistance in septic shock was significantly higher in the group of surviving patients compared with patients who died. According to their observations, one can postulate that lowering of cardiac output can reduce cardiac work in SIRS which is an important factor for patients with diminished cardiac reserve (i.e. ischemic heart disease, myocardial insufficiency). An additional factor with benefit in SIRS could be the amelioration of D_{O_2}/V_{O_2} quotient. Many studies demonstrated enhanced need of oxygen in SIRS. There is a discrepancy between D_{O_2} and oxygen consumption (V_{O_2}) described as a pathological dependency on oxygen consumption from oxygen delivery [14, 21–23]. Supranormal oxygen delivery by volume therapy, blood transfusion as well as administration of inotropic and vasodilating agents are one way of therapeutic intervention. Some authors are in doubt about this method because they could not measure positive results or they have seen negative effects [24–27]. Results presented in this study may favor the hypothesis of Vincent and van der Linden [28]. These authors express the opinion that enhancing oxygen extraction and not oxygen delivery would be the best to influence the pathologic supply dependency [28]. But we have to keep in mind that this hypothesis remains unproved till now [28]. One important aspect is that CVVH could eliminate some cytokines which are involved in microcirculatory dysregulation of SIRS and inhibition of mitochondrial oxidative pathway [29, 30].

CVVH after severe polytrauma leads to reduced activation of the humoral cascade as can be shown by lower complement levels. Decreased levels of myeloperoxidase in patients with severe polytrauma treated by CVVH indicate limited membrane injury, e.g. caused by diminished polymorphonuclear cell activation in SIRS patients under CVVH treatment. The relevance of activated polymophonuclear granulocytes in the development of acute respiratory distress syndrome and multiple organ failure has been clearly shown earlier [9, 31].

Significantly elevated levels of TNF-α were measured in 8 of 12 polytrauma patients treated by conventional method alone but only in 3 of 12 patients receiving hemofiltration. This result may reflect an attenuated systemic inflammatory reaction because TNF-α is a key mediator of SIRS. The role of removal of TNF-α by hemofiltration could be discussed because TNF-α was also determined in the ultrafiltrate of septic patients [4, 5]. TNF can be found in the serum as a complex of four molecules which cannot pass the filter. This effect is also described for IL-6. Tonnesen et al. [5] have demonstrated that IL-6 was not filtered through membrane despite elevated plasma levels. We also observed no changes for IL-6.

C-reactive protein was elevated during the first 3 days in patients undergoing hemofiltration. The synergistical effect of TNF-α, IL-1 and IL-6 on the synthesis of this acute phase protein cannot explain the results. They are lowered or unchanged. Moreover, we have to notice that C-reactive protein has no specific character and is enhanced by infections or other inflammatory stimuli. One cannot exclude that the polyacrylonitrile membrane used for hemofiltration could be responsible for this process [32].

Lipid peroxidation represents the effects at the level of the cell membrane. Malondialdehyde is a stable end product of lipid peroxidation and mostly derived from vascular endothelial cells. The significantly lowered levels of malondialdehyde in the patients undergoing hemofiltration after polytrauma provide evidence for reduced tissue injury by hemofiltration. The consequences could be a protection of secondary organ failure induced by SIRS and an amelioration of systemic hemodynamic situation.

Conclusion

Early treatment of polytrauma patients by CVVH leads to an attenuation of a hyperdynamic circulatory situation. Enhanced cardiac output and oxygen delivery were attenuated towards normal values while oxygen consumption remained constant. These effects are probably due to a reduced posttraumatic inflammatory response which is characterized by reduction of some inflammatory markers in

CVVH patients. Moreover, the significant reduction of plasma lipid products peroxidation gives evidence of reduced organ injury in polytrauma patients undergoing CVVH treatment.

References

1 Goris JA, Te Boekhorst TPA, Nuytinck JKS, Gimbrere JSF: Multiple organ failure. Arch Surg 1985;120:1109–1115.

2 Beal Al, Cerra FB: Multiple organ failure syndrome in the 1990s: Systemic inflammatory response and organ dysfunction. JAMA 1994;271:226–233.

3 Lowry SF: Cytokine mediators of immunity and inflammation. Arch Surg 1993;128:1235–1241.

4 Bellomo R, Tipping P, Boyce N: Continuous veno-venous hemofiltration with dialysis removes cytokines from the circulation of septic patients. Crit Care Med 1993;21:522–526.

5 Tonnesen E, Hansen MB, Höhndorf K, Diamant M, Bendtzen K, Wanscher M, Toft P: Cytokines in plasma and ultrafiltrate during continuous arteriovenous haemofiltration. Anaesth Intensive Care 1993;21:752–758.

6 Blinzler L, Zaar P, Leier M, Bürger R, Heuser D: Multiple organ failure: Therapeutic approach with continuous venovenous hemofiltration? (abstract). Clin Intensive Care 1994;5(suppl):25.

7 Hoffmann JN, Faist E, Deppisch R, Inthorn D: Septic ultrafiltrate enhances TNF-alpha production in endotoxin-stimulated human mononuclear cells and down-regulates lymphocyte function (abstract). Clin Intensive Care 1994;5(suppl):24.

8 Marzi I, Bauer C, Hower R, Bühren V: Leukocyte-endothelial cell interactions in the liver after hemorrhagic shock in the rat. Circ Shock 1993;40:105–114.

9 Carrico CJ, Meakins JL, Marshall JC, Fry D, Maier RV: Multiple organ failure syndrome. Arch Surg 1986;121:196–208.

10 Marzi I, Bühren V, Schüttler A, Trentz O: Value of superoxide dismutase for prevention of multiple organ failure after multiple trauma. J Trauma 1993;35:110–120.

11 Heinrich PC, Castell JV, Andus T: Interleukin-6 and the acute phase response. Biochem J 1990; 265:621–636.

12 Bone RC, Balk RA, Cerra FB, et al: American College of Chest Physicians Society Critical Care Medicine Consensus Conference: Definitions for sepsis and organ failure and guidelines for the use of innovative therapies in sepsis. Chest 1992;101:1644–1655.

13 Soni N, Fawcett WJ, Halliday FC: Beyond the lung: Oxygen delivery and tissue oxygenation. Anesthesia 1993;48:704–711.

14 Vincent JL: The relationship between oxygen demand, oxygen uptake, and oxygen supply. Intensive Care Med 1990;16:S145–S148.

15 Bagshaw ONT, Anaes FRC, Hutchinson A: Continuous arteriovenous haemofiltration and respiratory function in multiple organ systems failure. Intensive Care Med 1992;18:334–338.

16 Grazia F, Todor R, Scalea T: Continuous arteriovenous hemofiltration countercurrent dialysis in acute respiratory failure. J Trauma 1991;31:1277–1284.

17 Sold M: Hämofiltration beim Multiorganversagen. Anästh Intensivmed 1992;33:349–353.

18 Gomez A, Wang R, Unruh H, et al: Hemofiltration reverses left ventricular dysfunction during sepsis in dogs. Anesthesiology 1990;75:671–685.

19 Seefelder C, Matzek N, Rossi R: Polytrauma-Bewertungsskalen. Notfallmedizin 1988;14:317–328.

20 Groeneveld ABJ, Nauta JJP, Thijs LG: Peripheral vascular resistance in septic shock: Its relation to outcome. Intensive Care Med 1988;14:141–147.

21 Edwards JD: Oxygen delivery and consumption. Clin Intensive Care 1990;16:62–65.

22 Nightingale P: Practical points in the application of oxygen transport principles. Intensive Care Med 1990;16:S173–S177.

23 Shoemaker WC, Apple PL, Kram HB, et al: Prospective trial of supranormal values of survivors as therapeutic goals in high-risk surgical patients. Chest 1988;94:1176–1186.

24 Barone JE, Lowenfels AB: Maximization of oxygen delivery: A plea for moderation. J Trauma 1992;33:651–653.
25 Dantzker DR, Foresman B, Guitterez G: Oxygen supply and utilization relationship. A reevaluation. Am Rev Respir Dis 1991;143:675–679.
26 Oh TE, Bhatt SB: Oxygen delivery and oxygen consumption (correspondence). Anesthesia 1993;48: 731–732.
27 Rudolph T, Meier-Hellmann A, Hannemann L, Reinhart K: Oxygen transport pattern in surviving and nonsurviving septic patients treated to maximize oxygen delivery (abstract). Clin Intensive Care 1994;5(suppl):22.
28 Vincent JL, van der Linden P: Septic shock: Particular type of acute circulatory failure. Crit Care Med 1990;18:S70–S74.
29 Robertson FM, Offner JP, Ciceri DP, Becker WK, Pruitt BA: Detrimental hemodynamic effects of nitric oxide synthetase inhibition in septic shock. Arch Surg 1994;129:149–156.
30 Stadler J, Bentz B, Harbrecht BG, et al: Tumor necrosis factor-alpha inhibits hepatocyte mitochondrial respiration. Ann Surg 1992;216:539–546.
31 Swank DW, Moore SB: Roles of the neutrophil and other mediators in adult respiratory distress syndrome. Mayo Clin Proc 1989;64:1118–1132.
32 Hörl WH, Riegel W, Steinhauer C, Wanner C, Thaiss F, Bozkurt F, Haag M, Schollmeyer P: Granulocyte activation during hemodialysis. Clin Nephrol 1986;26(suppl 1):S30–S34.

Priv.-Doz. Dr. Werner Riegel, Medizinische Universitätsklinik IV,
D–66421 Homburg/Saar (Germany)

Sieberth HG, Stummvoll HK, Kierdorf H (eds): Continuous Extracorporeal Treatment in Multiple Organ Dysfunction Syndrome. Contrib Nephrol. Basel, Karger, 1995, vol 116, pp 62–75

Cytokine Kinetics (TNF-α, IL-1β, IL-6) during Continuous Hemofiltration: A Laboratory and Clinical Study

E.F.H. van Bommel[a], *C.J. Hesse*[a], *N.H.P.M. Jutte*[a], *R. Zietse*[a], *H.A. Bruining*[b], *W. Weimar*[a,1]

[a] Department of Internal Medicine I, Division of Nephrology, and
[b] Surgical Intensive Care Unit, University Hospital Rotterdam 'Dijkzigt', The Netherlands

The development of acute renal failure (ARF) in intensive care unit (ICU) patients, usually as part of multiple organ systems failure (MOSF), carries a poor prognosis [1, 2]. Although continuous hemo(dia)filtration (CHF) is now increasingly accepted as the preferred treatment modality for the critically ill patient with ARF, it is not clear whether its use has an impact on outcome of these patients [2].

Both experimental [3, 4] and uncontrolled clinical data [2, 5, 6] suggest that CHF may improve hemodynamics and gas exchange in patients with septic shock and (nonrenal) MOSF, independent of an impact on fluid balance. These beneficial effects are ascribed to the removal from the circulation of middle and large molecular weight (>500 daltons) inflammatory mediators, particularly cytokines, responsible for the cardiorespiratory and metabolic alterations leading to MOSF [7–9]. Indeed, (persistently) elevated levels of tumor necrosis factor-α (TNF-α) and interleukin (IL)-6 have been associated with progressive inflammatory illness and poor outcome in both septic and nonseptic shock [7, 8]. However, few and controversial data exist concerning the removal cytokines from the circulation with CHF [10, 11].

The aim of the present study was to investigate whether cytokines are removed from the circulation with CHF both under in vitro conditions and in

[1] We wish to thank A.M. Vendrik for technical assistance with the in vitro hemofiltration experiments.

vivo, i.e., in patients with ARF suffering from the systemic inflammatory response syndrome (SIRS) [12]. In addition, it was investigated whether these cytokines were sieved through or adsorbed to the hemofiltration membrane.

Materials and Methods

In vitro Conditions

CHF was conducted in vitro using an experimental set-up, consisting of a 'blood' reservoir containing Ringer's lactate with human albumin (15 g/l), a synthetic hemofilter with standard arterial, venous and ultrafiltrate tubings, and roller pumps (Rhône-Poulenc, France) to regulate blood and ultrafiltrate rates. The venous outflow and ultrafiltrate were returned to the blood reservoir, thus minimizing changes in blood volume during the experiment. Polyacrylonitrile (AN 69-HF, 0.60 m^2 Hospal Ltd, France), polyamide (PAM) (FH66-D, 0.60 m^2; Gambro BV, Sweden) and polysulfone (PS) (BL-624, 1.0 m^2; Sorin Biomedica, Germany) hollow-fiber hemofilters were used. In consecutive experiments using a single mediator, recombinant TNF-α (16.5 kD) and IL-1β (16.5 kD), respectively, were added to the blood reservoir and allowed to equilibrate for 15 min. The pumps were calibrated to deliver a blood flow (Q_b) of 150 ml/min and an ultrafiltrate flow (Q_{uf}) of 30 ml/min. During initial experiments, samples (2 ml) were taken from the arterial, venous and ultrafiltrate line at 15, 60, and 120 min. At 0 and 150 min, samples were taken from the reservoir. Because these results suggested that the sieving properties of cytokines were masked by their initial adsorption and that the binding capacity of the AN 69 membrane was not saturated for TNF-α, we subsequently performed CHF with the AN 69 membrane during a 24-hour period, again using a single mediator (TNF-α, IL-1β, and IL-6 (22,0 kD), respectively) in each experiment. Samples were taken immediately prior to and at 5, 10, 15, 30, 60, 120, 240, 360, 480, 720 and 1,440 min, respectively, after the start of CHF. The amount of cytokine left unbound was calculated from the 'blood' volume (corrected for sample collection) and final cytokine concentration. The amount of cytokine bound is equal to the amount added initially minus the amount remaining unbound. Samples were drawn in ethylenediamine (EDTA) tubes and immediately stored at –70 °C until assays were performed.

In vivo Conditions

Patients

Twenty critically ill patients with ARF who were treated by continuous arteriovenous hemodiafiltration (CAVH-D) in a single surgical ICU were included in a prospective study during the period August 1, 1992–January 31, 1994. All patients suffered from SIRS, as manifested by two or more of the following conditions: temperature > 38 °C or < 36 °C; heart rate > 90 beats/min; respiratory rate > 20 breaths/min or PaO_2 < 4.3 kPa (< 32 Torr); WBC > 12,000 cells/mm^3, < 4,000 cells/mm^3, or > 10% immature (band) forms [12]. All patients required mechanical ventilation and inotropic support prior to the institution of CAVH-D. The following information was obtained: age, gender, underlying disease(s) and surgical procedures performed, presence of systemic infection, severity of illness as calculated from the

Acute Physiology and Chronic Health Evaluation (Apache II) score [13] and the number of failed organ systems [12, 13], outcome and cause of death. Cardiorespiratory and acid-base variables were monitored prior to the start of CAVH-D and 1, 4, and 24 h, respectively, after the start of treatment.

Continuous Arteriovenous Hemodiafiltration

CAVH-D was performed using a hollow-fiber polyacrylonitrile hemofilter (AN 69-HF) connected to the arterial aand venous vascular access devices in a pumpless circuit. Vascular access was by means of large-bore femoral catheters (Medcomp, Harleysville, Pa., USA) (n = 17) or by means of a Scribner shunt (Quinton, N.J., USA) (n = 3). Warmed bicarbonate-buffered dialysis fluid (SH 44-HEP; Schiwa GmbH, Glandorf, Germany) was pumped countercurrent to blood through the ultrafiltrate compartment of the filter at a rate of 1 liter/h with a calibrated, volumetric pump (3M, AVI Inc., St. Paul, Minn., USA). Lactate-based replacement fluid was given as clinically indicated, administered into the arterial limb of the extracorporeal circuit. The extracorporeal circuit was anticoagulated with heparin (500–700 U/h) unless there was severe coagulopathy or thrombocytopenia ($<60 \cdot 10^9$/l). The filters were changed when there was objective evidence of inadequate filter function (i.e., a sustained reduction in ultrafiltrate volume (<300 ml/h) or evidence of filter clotting. Net ultrafiltration rate at each time point was measured from timed volumetric collections (Q_{uf} (ml/min)).

Sample Collection

Both blood and ultrafiltrate levels of the cytokines TNF-α, IL-1β, and IL-6 were measured in these patients (n = 20). Blood samples were taken from a port on the arterial tubing prior to the infusion of predilution fluid and, as such, represent the patient's systemic levels. Simultaneously, samples were drawn from the ultrafiltrate port. To assess cytokine sieving properties (i.e., no dilution of ultrafiltrate by dialysis fluid), the dialysate line was clamped and dialysate pump stopped 15 min prior to collection so as to obtain 'pure' ultrafiltrate, i.e., *convective* solute transfer occurs only (CAVH). To assess cytokine mass transfer across the membrane, clearances of cytokines from blood and the impact of prolonged blood-membrane interaction on these characteristics, blood samples were also taken from a port immediately postfilter in 10 of these patients. To calculate the blood flow pre- ($Q_{b_{in}}$) and postfilter ($Q_{b_{out}}$), hematocrit (Hct) levels were measured simultaneously at the filter inlet and outlet (Hct_{in}, Hct_{out}) in these patients (n = 10). For each patient, samples were taken immediately prior to and 1, 4 and 24 h, respectively, after the initiation of CAVH-D. Postfilter values were corrected for hemoconcentration as a result of ultrafiltration. The 'age' of the filter was recorded at each sampling time point because filter replacement was required within 24 h in some patients as a result of filter clotting. Samples were placed in EDTA tubes and held on ice until centrifugation. The plasma was removed and stored at −70 °C until assays were performed. Ultrafiltrate samples were drawn in EDTA tubes and held on ice until storage at −70 °C.

Calculations. The following formulae were used [14]: Blood flow: $Q_{b_{in}} = (Q_{uf} \cdot Hct_{out})/Hct_{out} - Hct_{in}$; $Q_{b_{out}} = Q_{b_{in}} - Q_{uf}$ (ml/min); sieving coefficient: $S = 2 \cdot C_{uf}/C_{in} + C_{out}$; clearance (filtrate): $K_{uf} = C_{uf} \cdot Q_{uf}/C_{in}$ (ml/min); clearance (blood): $K_b = ((Q_{b_{in}} \cdot C_{in}) - (Q_{b_{out}} \cdot C_{out}))/C_{in}$ (ml/min); mass transfer (J) across the membrane from blood: $J_b = (Q_{b_{in}} \cdot C_{in}) - (Q_{b_{out}} \cdot C_{out})$ (pg/min), and into ultrafiltrate: $J_{uf} = Q_{uf} \cdot C_{uf}$ (pg/min); mass balance: $J_{bin} - J_{bout} = J_{uf}$.

Assay Procedures. Blood and ultrafiltrate concentration of cytokines were measured by commercially available enzyme-linked immunosorbent assays (ELISA) and performed according to the manufacturer's instructions. TNF-α concentrations were measured with a one-step 'sandwich' type immunoassay (Immunotech SA, Marseille, France). The lower detection limit of TNF-α was 7.8 pg/ml. The coefficient of variation was < 10% in the range of 115–1,000 pg/ml. No cross-reactivity appears with TNF-β, IL-1β, IL-6, or the p55 and p75 forms of the TNF receptor. IL-1β concentrations were measured using a two-step immunoassay (Immunotech SA). Analytical sensitivity of the assay was 5 pg/ml. Coefficient of variation was <8% in the range of 50–1,000 pg/ml. No cross-reactivity occurs with IL-α, IL-6, or TNF-α. IL-6 concentrations were measured with a three-step immunoassay, using the human (hu) IL-6 ELISA compact kit (CLB, Amsterdam, The Netherlands). Analytical sensitivity of the assay was 5 pg/ml. Coefficient of variation was <10% in the range of 6.5–400 pg/ml. No cross-reactivity occurs with either IL-1β or TNF-α. Samples were assayed in duplicate and all cytokine levels assayed in the same run. Cytokine concentrations, where necessary, were corrected for dilution. The mean of the duplicate measurements is reported at each time period.

Statistical Analysis. Data are expressed as mean ± SEM or percentage, where appropriate. The Wilcoxon signed-rank test was used to analyze nonparametric data. Repeated measures ANOVA test was used to analyze normally distributed data. Statistical significance was accepted at a $p < 0.05$ level.

Results

In vitro Conditions

Different Membranes

Significant amounts of TNF-α (3.5 µg) disappeared from the blood compartment with the AN 69 membrane (fig. 1) with only minimal amounts detectable in the ultrafiltrate. The only plausible explanation for the observed mass balance error in the closed-loop recirculation model is that TNF-α is bound to the membrane. No appreciable adsorption of TNF-α was noted for the PS (0.15 µg) and PAM (0.3 µg) membrane (fig. 1), steady-state blood levels occurring after 15 min. There was no appreciable sieving of TNF-α during these short-term hemofiltration experiments (fig. 2). However, sieving properties of TNF-α with the AN 69 membrane seemed to increase over time with a concomitant increase of convective clearance (K_{uf} 0 ml/min (15 min); 0.8 ml/min (60 min); 2.9 ml/min (120 min)). Only minimal convective clearance occurred with the two other membranes (K_{uf} (PS) 0.96 ml/min; (PAM) 0.24 ml/min at 120 min). The adsorption capacity of all three membranes for IL-1β was less compared to that seen for TNF-α (AN 69 1.0 µg; PAM 0.03 µg; PS 0.06 µg). However, significant sieving of IL-1β was seen with all three membranes, again with increasing sieving properties

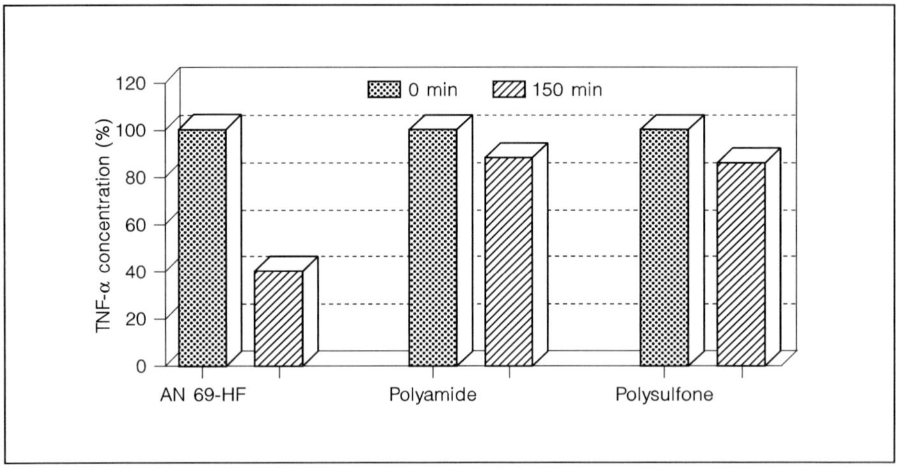

Fig. 1. TNF-α elimination (monomeric form; 16.5 kD) over time during in vitro hemofiltration with three different membranes. Both the ultrafiltrate and venous outflow were returned to the 'blood' reservoir; therefore, mass lost from 'blood' could only be due to adsorption. Results are expressed as percent of total TNF-α measured in the blood reservoir before the start of recirculation. Only the AN 69 membrane exhibited appreciable adsorption.

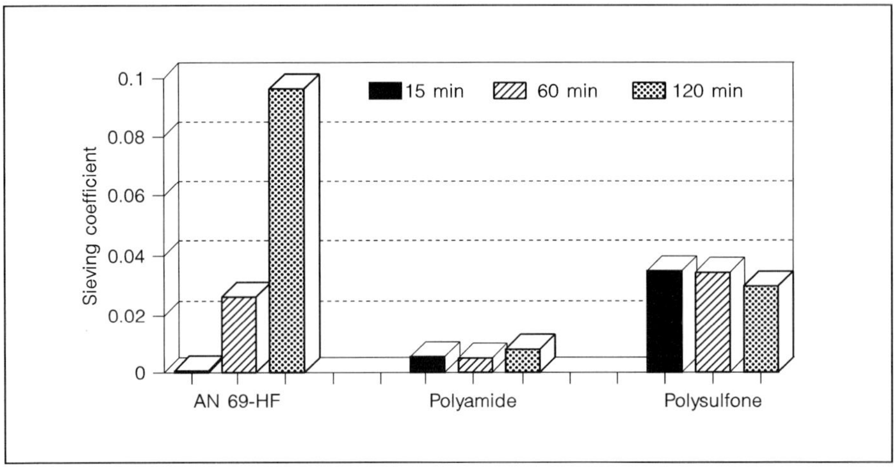

Fig. 2. TNF-α sieving (monomeric form; 16.5 kD) over time during in vitro hemofiltration with three different membranes. No appreciable sieving of TNF-α was observed during the first 60 min of hemofiltration; thereafter, sieving properties of the AN 69 membrane increased with the sieving coefficient of TNF-α approaching 0.1 (10%).

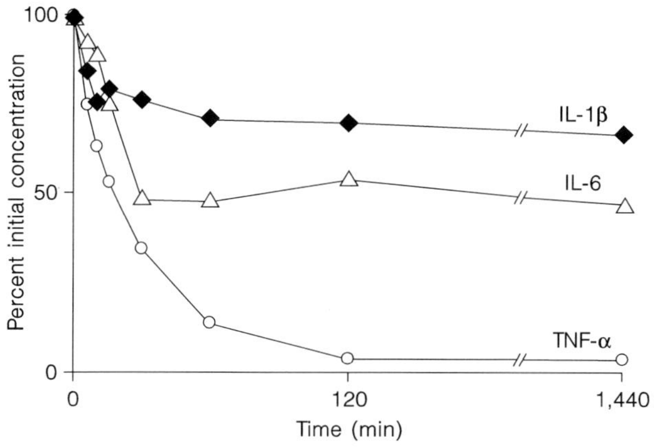

Fig. 3. Arterial (i.e., prefilter) cytokine levels (TNF-α, IL-1β, IL-6) over time during continuous hemofiltration. Both the ultrafiltrate and venous outflow were returned to the 'blood' reservoir; therefore, mass lost from 'blood' could only be due to adsorption. A single mediator was added in each experiment. Results are expressed as percent of total cytokine measured in the 'blood' before the start of recirculation. For the purpose of clarity, steady-state cytokine values between 240 and 1,440 min are not shown.

over time with the AN 69 membrane (S (AN 69) 0.96; (PS) 0.67; (PAM) 0.62 at 120 min). This resulted in significant convective clearances of IL-1β (K_{uf} (AN 69) 28.8 ml/min; (PS) 20.1 ml/min; (PAM) 16.5 ml/min at 120 min) with all three membranes.

Twenty-Four Hour Hemofiltration with the AN 69 Membrane

During the 24-hour hemofiltration experiments with the AN 69 membrane, the adsorption capacity of the membrane became saturated for all three cytokines, as indicated by steady-state arterial (i.e., prefilter) cytokine levels (fig. 3). Calculated adsorption capacity of the membrane for TNF-α was 4.5 μg, IL-1β 1.6 μg, and for IL-6 1.9 μg. Sieving capacities of the different cytokines were masked by their initial adsorption. When the binding capacity of the membrane became saturated, the sieving coefficient of all cytokines increased significantly (TNF-α 0.01 ± 0.003 to 1.0 ± 0.01; IL-1β 0.12 ± 0.01 to 1.0 ± 0.14; IL-6 0.01 ± 0.004 to 0.32 ± 0.031; $p < 0.05$ (n = 3, mean ± SEM)).

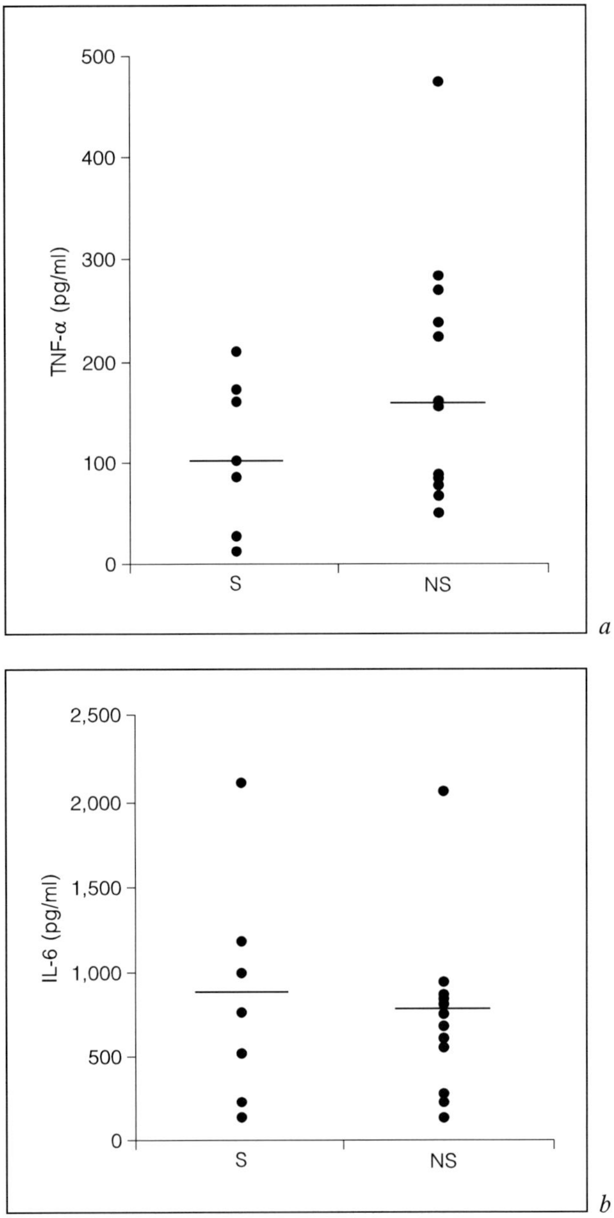

Fig. 4. Individual plasma TNF-α (*a*) and IL-6 (*b*) concentrations in surviving (S) and nonsurviving (NS) patients. TNF-α levels were higher in nonsurviving patients ($p < 0.05$).

Table 1. Cardiorespiratory and acid-base variables prior to and during CAVH-D treatment

Variable	Time, h			
	pre	1	4	24
MAP, mm Hg	73±1.8	76±2.0	84±2.1	78±1.8
CO, l/min	5.8±0.5	6.1±0.4	6.4±0.4	6.0±0.3
HR, beats/min	108±2.3	106±2.1	101±1.8	104±1.7
SVR, $dyn \cdot s \cdot cm^{-5}$	724±82.1	732±61.2	884±121.1	801±112.3
MPAP, mm Hg	27±2.5	29±2.1	26±2.0	28±0.4
PCWP, mm Hg	15±1.1	13±0.8	15±1.1	14±0.8
PaO_2/FiO_2 ratio	2.3±0.3	2.4±0.4	2.6±0.3	2.6±0.4
pH	7.36±0.3	7.38±0.3	7.40±0.4	7.39±0.3
Bicarbonate, mmol/l	16.2±0.7	17.4±1.6	17.5±0.9	18.9±0.4*

MAP = Mean arterial pressure; CO = cardiac output; HR = heart rate; SVR = systemic vascular resistance; MPAP = mean pulmonary arterial pressure; PCWP = pulmonary capillary wedge pressure.

All patients required artificial ventilation and inotropic support (n = 20). Daily ultrafiltrate volume 12.6 ± 0.42 liters/day; net ultrafiltration was 3.2 ± 0.1 liters/day. Net 24-hour fluid balance was plus 1,208 ± 564 ml. Results are expressed as mean ± SEM. * $p < 0.05$ vs. pretreatment value.

In vivo Conditions

Patients

Mean age of the patients was 54 ± 3.8 years (range 17–83) with a male predominance (75%). The mean pretreatment Apache II score was 27 ± 0.7 (range 19–34) and the mean number of organ system failures was 3.3 ± 0.3 (range 1–6). Mean arterial pressure (MAP) upon initiation of CAVH-D was 73 ± 1.8 mm Hg (range 55–86). Eight patients (40%) survived to leave the ICU, of whom 7 (35%) survived to leave hospital. Nine patients (69%) died in refractory septic shock complicated by MOSF, 3 patients (23%) died from irreversible MOSF following multitrauma, and 1 patient (8%) died from myocardial infarction. All patients had detectable TNF-α and IL-6 plasma levels. The plasma TNF-α levels in nonsurvivors were higher than in survivors (fig. 4a). No difference in plasma IL-6 levels was observed between survivors and nonsurvivors (fig. 4b). None of the patients had detectable IL-1β plasma levels. Hemodynamic, gas exchange, and acid-base variables prior to and during 24 h of CAVH-D treatment are shown in table 1. There was a nonsignificant trend to improved hemodynamics and gas exchange after 4 h of treatment, which could at least in part be explained by fluid removal. In addition, a significant increase in serum bicarbonate levels was observed after 24 h of treatment.

Ultrafiltrate Clearances

Mean ultrafiltrate volume during CAVH-D was 12.6 liters/day (range 7.8–14.8). Mean blood flow through the filter was 92 ± 3.8 ml/min. Overall, the mean plasma TNF-α level upon initiation of CAVH-D was 153 ± 7.3 pg/ml (range 26–475). TNF-α could be detected in only 17 of 60 (28%) ultrafiltrate specimens (mean TNF-α concentration 8.9 ± 0.8 pg/ml; range 2.1–48). Mean daily mass transfer of TNF-α into the ultrafiltrate was 0.03 ± 0.01 μg/day (range 0.0–0.2). Mean daily ultrafiltrate clearance of TNF-α was 0.1 ± 0.02 liter/day. No significant change in the plasma TNF-α level was observed over time following the initiation of CAVH-D (pretreatment TNF-α 153 ± 7.3 vs. 24 h TNF-α 170 ± 41.7 pg/ml; NS). The mean IL-6 concentration upon initiation of CAVH-D was 795 ± 163.2 pg/ml (range 265–2,097). Significant amounts of IL-6 were detected in all 60 ultrafiltrate specimens (mean IL-6 concentration 304 ± 128.6 pg/ml; range 34–1,294). Mean mass transfer of IL-6 into the ultrafiltrate was 3.4 ± 0.3 μg/day (range 1.4–6.1). Mean daily ultrafiltrate clearance was 3.9 ± 1.8 liters/day. There was a significant decrease of plasma IL-6 level over time following the initiation of CAVH-D (pretreatment IL-6 level 796 ± 163.2 vs. 24 h 548 ± 82.2 pg/ml; $p < 0.05$).

Filter Duration

No differences were found between the mean sieving coefficient of TNF-α, when correlated to the age of the filter (S 0.02 ± 0.01 (1–2 h), 0.01 ± 0.01 (4–6 h), 0.02 ± 0.01 (20–24 h); NS). Postfilter TNF-α levels, corrected for hemoconcentration, were significantly lower with filters used for 1–2 h (prefilter TNF-α 152 ± 43.2 vs. postfilter TNF-α 126 ± 36.7 pg/ml; $p < 0.05$). No difference between pre- and postfilter levels of TNF-α were observed with filters used for 4–6 h (178 ± 47.8 vs. 184 ± 51.2 pg/ml; NS). Significant clearance of TNF-α from blood occurred with filters used for ≤6 h (fig. 5a). Mass transfer of TNF-α from blood with filter duration ≤6 h was greater than that appearing in the ultrafiltrate (J_b 2.7 ± 0.8 vs. J_{uf} 0.03 ± 0.01 ng/min (1–2 h); J_b 0.5 ± 0.7 vs. J_{uf} 0.02 ± 0.01 ng/min (4–6 h); $p < 0.001$), suggesting that membrane adsorption occurs. However, with prolonged filter use, postfilter values were significantly higher than prefilter values (prefilter TNF-α 159 ± 22.1 vs. postfilter TNF-α 233 ± 58.6 pg/ml (20–24 h); $p < 0.05$). Negative mass balance (–4.3 ± 1.6 ng/min) and blood clearance of TNF-α with filters used for ≥20 h (fig. 5a), suggest that either desorption of increased generation of TNF-α occurs. No differences in the mean sieving coefficient of IL-6 were found, when correlated to filter duration (S 0.28 ± 0.06

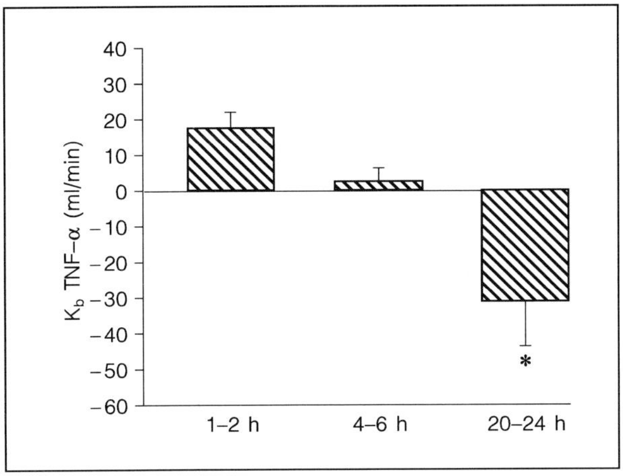

a

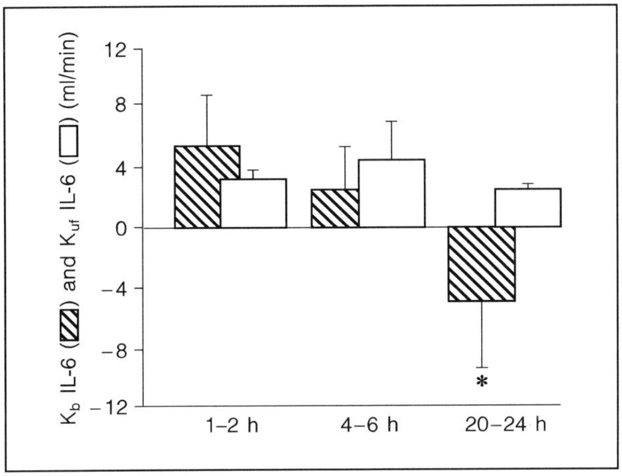

b

Fig. 5. Clearance of TNF-α (*a*) and IL-6 (*b*) from blood (K_b) and ultrafiltrate (K_{uf}), correlated to filter duration, during continuous hemofiltration with the AN 69 membrane. K_b of TNF-α and IL-6 declined following prolonged filter use and became negative over time. No appreciable ultrafiltration of TNF-α (K_{uf} <0.2 ml/min) occured. * $p < 0.05$ vs. 1–2 h.

(1–2 h), 0.30 ± 0.08 (4–6 h), 0.24 ± 0.02 (20–24 h); NS). Postfilter levels of IL-6, corrected for hemoconcentration, were lower with filters used for 1–2 h (prefilter IL-6 level 733 ± 160 vs. 628 ± 100 pg/ml; $p < 0.05$). Pre- and postfilter levels of IL-6 did not differ significantly with filters used for ≥4 h (699 ± 114 vs. 692 ± 153 pg/ml (4–6 h), and 665 ± 89 vs. 735 ± 106 pg/ml (20–24 h); NS). Significant blood clearance of IL-6 occurred with filters used for ≤6 h, but became negative with prolonged filter use (fig. 5b). Mass transfer of IL-6 from blood did not differ significantly from that appearing in the ultrafiltrate for any given filter duration.

Discussion

The present study demonstrates that circulating cytokines can be removed during CHF. During the first period of in vitro CHF, significant amounts of TNF-α disappeared from the blood compartment with only minimal amounts detectable in the ultrafiltrate. The only plausible explanation for the observed mass balance error in the closed-loop recirculation model is that TNF-α is bound to the membrane. Results also suggested that the sieving property of the monomeric form of TNF-α was masked by the large capacity of the AN 69 membrane to bind TNF-α. When the membrane was saturated, mass transfer of TNF-α into the ultrafiltrate was maximal (i.e., the S of TNF-α was 1.0). These findings are in accordance with those of Leypoldt et al. [15] who found that initial sieving of model macromolecules (i.e., polydisperse DEAE dextran) during hemofiltration were masked by their rapid adsorption to the AN 69 membrane. We also noted significant adsorption of this cytokine to the AN 69 membrane in vivo. However, the sieving capacity of TNF-α in vivo was low, irrespective of filter duration. Other investigators also detected no [10, 16] or only small amounts of TNF-α [17, 18] in ultradiafiltrate samples during CHF. One explanation for the low sieving capacity of TNF-α in vivo may be that it exists in its bioactive form as a trimer (52 kD) [19], thereby exceeding the molecular cut-off point of the membrane. Another explanation may be that TNF-α is rapidly bound to the soluble TNF receptor (sTNFR) [20]. When plasma levels of TNF-α are <500 pg/ml, there is a positive correlation between sTNFR and TNF-α levels. However, when plasma levels of TNF-α exceed 500 pg/ml, levels of sTNFR may not rise proportionally [20]. With very high TNF-α levels, therefore, it may be that a greater proportion of TNF-α circulates unbound, thus being potentially removed via the ultrafiltrate. This might also explain the difference in sieving characteristics of TNF-α between this and other studies (serum TNF-α levels varying from 60 to 153 pg/ml), in which little or no sieving of TNF-α was observed [10, 16, 18], and the study of Bellomo et al. [11] (mean serum TNF-α level 765 pg/ml), in which significant amounts of TNF-α were detected in 69.5% of ultradiafiltrate samples during CVVHD in septic patients. In the present study, significant removal of IL-6 via the ultrafiltrate was demonstrated in all patients. In addition, serum IL-6 levels decreased over time during CAVH-D. Gueugniaud et al. [20] also detected significant amounts of IL-6 in ultradiafiltrate samples of 4 severely burned patients and decreasing serum levels of this cytokine during CVVHD. Our in vitro data suggest that significant removal of IL-1β via the ultrafiltrate can be expected. Although we could not confirm our in vitro data because of lack of elevated serum IL-1β levels in our patients, others have noted significant amounts of this cytokine in the ultradiafiltrate of septic patients [11]. However, because of its transient appearance [7, 8], significant IL-1β clearance will probably occur only when CHF is instituted during the early phase of (non)septic shock.

The binding capacity of the AN 69 membrane for IL-1β and IL-6 was significantly less compared to that for TNF-α. In addition to these cytokine-dependent differences, our in vitro data suggest that synthetic membranes differ in adsorption capacity. This might explain why Byrick et al. [10] did not find removal of TNF-α from the circulation during CAVH-D without ultrafiltration with the PS membrane, irrespective of filter duration, in a patient with rhabdomyolysis. The affinity of synthetic membranes for cytokines may reside in their hydrophobic surfaces [21]. One important factor explaining the comparatively high adsorption capacity of the AN 69 membrane may be its electrical charge. For example, the anaphylatoxins C3a (9.0 kD) and C5a (11.2 kD) have shown to display high affinity for the AN 69 membrane, presumably because of electrostatic interactions [21].

Adsorption of cytokines as well as other proinflammatory peptides may be beneficial because of their removal from the circulation. Conversely, however, a high concentration of these proteins locally on the membrane surface may elicit an inflammatory response by stimulating adherent neutrophils and monocytes. We and others [18] observed higher postfilter TNF-α values, corrected for hemoconcentration, following prolonged filter use. An increase in IL-1β levels, although not corrected for hemoconcentration, was also observed after 4 h of CVVHD treatment in septic patients [11]. In addition, Kapan et al. [22] observed significantly higher postfilter values of another important proinflammatory peptide, i.e. C5a, corrected for hemoconcentration, during CHF in ARF patients. These results may be explained either by desorption of these peptides from the membrane [15] or by increased generation. Recent data indicate that blood-membrane interaction in chronic hemodialysis patients does not directly lead to increased cytokine production, but results in priming of mononuclear cells for enhanced release of cytokines following secondary stimulation, e.g. LPS fragments from contaminated dialysate [23]. Therefore, it is possible that the immunologic response to (prolonged) blood-membrane interaction during CHF in critically ill patients suffering from sepsis or SIRS is significantly different compared to that resulting from short-term blood-membrane interaction in uremic but otherwise healthy patients. However, despite this potentially adverse effect of prolonged blood-membrane interaction in critically ill patients, it does not seem to have any clinical significance. On the contrary, (improved) hemodynamic stability (table 1) is uniformly found during CHF [1, 2]. It may be that the positive effect of an unselective removal of both identified (e.g., cytokines, anaphylatoxins) and as-yet-to-be-identified vasoactive mediators from the circulation outweighs the possible negative effect of prolonged blood-membrane interaction during CHF. This would also explain the correlation between the ultrafiltrate volume and hemodynamic changes, recovery time of diuresis and survival, which has been found by some investigators in animal and retrospective, clinical studies [1, 2, 24].

In conclusion, the present study suggests that cytokines can be removed from the circulation with CHF, either by their adsorption to the membrane (i.e., TNF-α) or by their convective removal (i.e., IL-1β and IL-6, unbound TNF-α?) into the ultrafiltrate. In addition to these cytokine-dependent differences, synthetic membranes seem to differ in adsorption and sieving capacities, which may be of clinical relevance. However, prolonged blood-membrane interaction seemingly affects adversely cytokine clearances, possibly by desorption and/or increased generation. If removal of cytokines is of primary concern, high ultrafiltrate volumes and frequent change of filters are needed to extract significant amounts of cytokines from the circulation. However, further research is warranted to establish whether this will have an impact on the outcome of human septic and nonseptic shock. Until such studies are performed, the rationale for the use of CHF in critically ill patients in the absence of conventional indications for dialytic support remains unproven.

References

1 Van Bommel EFH, Leunissen KLM, Weimar W: Continuous renal replacement therapy for the critically ill: An update. J Intensive Care Med 1994;9:265–280.

2 Grootendorst AF, van Bommel EFH: The role of hemofiltration in the critically ill intensive care unit patient: Present and future. Blood Purif 1993;11:209–223.

3 Gomez A, Wang R, Unruh H, Light RB, Bose D, Chan T, Correa E, Mink S: Hemofiltration reverse left ventricular dysfunction during sepsis in dogs. Anaesthesiology 1990;73:671–685.

4 Grootendorst AF, van Bommel EFH, van der Hoven B, van Leengoed LAMG, van Osta ALM: High volume hemofiltration improves hemodynamics of endotoxin-induced shock in the pig. J Crit Care 1992;7:67–75.

5 Gotloib L, Barzilay E, Shustak A, Lev A: Sequential hemofiltration in nonoliguric high capillary permeability pulmonary edema of severe sepsis. The artificial kidney as an artificial endocrine lung. Resuscitation 1986;13:997–1000.

6 Garzia F, Todor R, Scalea T: Continuous arteriovenous hemofiltration-countercurrent dialysis in acute respiratory failure. J Trauma 1991;31:1277–1285.

7 Calandra T, Baumgartner J-D, Grau GE, Wu M-M, Lambert P-H, Schellekens J, Verhoef J, Glauser MP: Prognostic values of tumor necrosis factor/cachectin, interleukin-1, interferon-α, and interferon-γ in the serum of patients with septic shock. J Infect Dis 1990;161:982–987.

8 Pinsky MR, Vincent J-L, Deviere J, Alegre M, Kahn RJ, Dupont E: Serum cytokine levels in human septic shock. Relation to multiple-system organ failure and mortality. Chest 1993;103:565–575.

9 Calandra T, Gerain J, Heumann D, Baumgartner J-D, Glauser MP: High circulating levels of interleukin-6 in patients with septic shock: Evolution during sepsis, prognostic value, and interplay with other cytokines. Am J Med 1991;91:23–29.

10 Byrick RJ, Goldstein MB, Wong PY: Increased plasma tumor necrosis concentration in severe rhabdomyolysis is not reduced by continuous arteriovenous hemodialysis. Crit Care Med 1992;20: 1483–1486.
11 Bellomo R, Tipping P, Boyce N: Continuous veno-venous hemofiltration with diaylsis removes cytokines from the circulation of septic patients. Crit Care Med 1993;21:522–526.
12 American College of Chest Physicians/Soceity of Critical Care Medicine Consensus Conference: Definitions for sepsis and organ failure and guidelines for the use of innovative therapies in sepsis. Crit Care Med 1992;20:864–874.
13 Knaus WA, Draper EA, Wagner DP, Zimmerman JE: Apache II: A severity of disease classification system. Crit Care Med 1985;13:818–929.
14 Henderson LW: Biophysis of ultrafiltration and hemofiltration; in Drukker W, Parsons FM, Maher JF (eds): Replacement of Renal Function by Dialysis. Boston, Nijhof, 1986, pp 242–264.
15 Leypoldt JK, Gilson JF, Blindauer KM, Cheung AK: Macromolecule adsorption to hemodialysis membranes depends on molecular size. Blood Purif 1992;10:53–60.
16 Elliott D, Wiles CE, Reynolds HN: Removal of cytokines in septic patients using continuous veno-venous hemodiafiltration (letter). Crit Care Med 1994;22:718–719.
17 Cottrell AC, Metha RL: Cytokine kinetics in septic ARF patients on continuous arteriovenous hemodiafiltration (abstract). J Am Soc Nephrol 1992;3:42.
18 Smith RA, Baglioni C: The active form of tumour necrosis factor is a trimer. J Biol Chem 1987;262: 6951–6954.
19 Moldawer LL: Interleukin-1, TNF-α and their naturally occurring antagonists in sepsis. Blood Purif 1993;11:128–133.
20 Gueugniaud P-Y, Bertin-Maghit M, Hirschauer C, Petit P: Removal of cytokines in septic patients using continuous venovenous hemofiltration (letter). Crit Care Med 1994;22:717.
21 Cheung AK, Chenoweth DE, Otsuka D, Henderson LW: Compartmental distribution of complement activation products in artifical kidneys. Kidney Int 1986;30:74–80.
22 Kaplan AA, Toueg S, Kennedy TL: Complement kinetics during continuous arteriovenous hemofiltration: Studies with a new polysulphone hemofilter. Blood Purif 1988;6:27–36.
23 Dinarello CA: Cytokines: Agent's provocators in hemodialysis? Kidney Int 1992;41:683–694.
24 Van Bommel EFH, Grootendorst AF, van Leengoed LAMG: The influence of high volume hemofiltration on hemodynamics in porcine endotoxic shock. Blood Purif 1992;10:88–89.

E.F.H. van Bommel, MD, Department of Internal Medicine, Division of Nephrology,
Drechtsteden Hospital, Van der Steenhovenplein 1, NL–3317 NM Dordrecht (The Netherlands)

Sieberth HG, Stummvoll HK, Kierdorf H (eds): Continuous Extracorporeal Treatment in Multiple Organ Dysfunction Syndrome. Contrib Nephrol. Basel, Karger, 1995, vol 116, pp 76–79

Hemofiltration in Human Sepsis: Evidence for Elimination of Immunomodulatory Substances

Johannes N. Hoffmann, Eugen Faist, Reinhold Deppisch, Wolfgang H. Hartl, Dietrich Inthorn

Department of Surgery, Klinikum Grosshadern, Ludwig Maximilian University, Munich, Germany

Continuous hemofiltration (HF) is widely used for renal replacement therapy in patients who develop kidney failure as part of a multiple organ dysfunction syndrome (MODS). In these patients we demonstrated a correlation between the daily amount of ultrafiltrate and the survival rate [1]. HF may also improve specific organ function in animal sepsis models [2]. Therefore, it has been suggested that HF may eliminate toxic mediators of MODS [3] although this issue remains controversial. The present study examined whether HF can activate or eliminate established mediators in patients with septic MODS. Because the exact nature of all factors removed into the ultrafiltrate remains unclear, it appeared appropriate to evaluate the biological effect of ultrafiltrate by exposing several white blood cell subfractions to ultrafiltrate in vitro.

Materials and Methods

Continuous isovolemic venovenous HF was performed using a fiber hemofilter (FH 66; Gambro AB, Hechingen, FRG) and a flow-controlled roller pump (AK 10; Gambro AB, Hechingen, FRG) with a filtration rate of 2 liters/min in 16 patients with septic MODS (Elebute [4] score points ≥ 20) and in 5 healthy volunteers. Fluid replacement was achieved via a digital balance control system by a balanced and isotonic electrolyte solution (HF 23; Fresenius, Bad Homburg, FRG). Low-dose heparinization was performed to prevent filter clotting by adding 300 IU/h heparin into the afferent line of the hemofilter. Pre- and postfilter and ultrafiltrate concentrations of cytokines (IL-6, TNF-α), and of complement compounds (C3a, C5a, C5b-9) were measured shortly after the beginning of HF (t_0), 60 min later

Table 1. TNF and IL-6 release of human monocytes and IL-2 and IL-6 release of lymphocytes in vitro

	Septic patients (n = 16)			Healthy volunteers (n = 5)		
	ultrafiltrate at t_0	ultrafiltrate at t_{60}	NaCl controls	ultrafiltrate at t_0	ultrafiltrate at t_{60}	NaCl controls
Monocytes						
TNF, U/ml	23.5±3.9*	25.0±5.8*	16.3±4.1	18.0±4.9	22.3±4.6	21.4±4.0
IL-6, U/ml	709±191**	571±149**	1,568±331	800±231	918±209	1,079±364
Lymphocytes						
IL-2, U/ml	0.35±0.07**	n.d.	0.66±0.07	0.48±0.18	n.d.	0.41±0.12
IL-6, U/ml	4,295±896*	3,093±741**	6,496±873	5,879±2,670	5,316±1,833	5,433±1,497

Cultures were costimulated with endotoxin (1 µg/ml) and PHA (2.5 µg/ml), respectively. Cells were incubated with septic or healthy ultrafiltrate collected at t_0 and t_{60}. Ultrafiltrate and corresponding NaCl incubations were always performed in the same cell preparation.

* $p < 0.01$ vs. NaCl; ** $p < 0.001$ vs. NaCl; n.d. = not determined.

(t_{60}) and 240 min later (t_{240}). Standard preparations [5] of healthy monocytes (Mo) and lymphocytes (Ly) were incubated with ultrafiltrate and stimulated with endotoxin and phytohemagglutinin (PHA), respectively. Also, PHA-induced peripheral blood mononuclear cell (PBMC) proliferation was studied. Cell function was determined by measuring concentrations of several cytokines released in the supernatants and by PHA (0.5 µg/ml)-induced ^{3}H-thymidine uptake during 4 h measured by a β-counter. Isotonic saline solution (NaCl) served as control. For dilution experiments, PHA (0.5 µg/ml)-stimulated PBMC were incubated with ultrafiltrate at different final concentrations (0, 6.25, 12, 25, and 50%). Assay volume was kept constant by adding NaCl. All data were expressed as mean ± SEM. The differences between the incubation procedures and between means of different time points were compared with the Student t test.

Results

HF showed no signs of mediator activation, since the most sensitive parameter (pre/postfilter C5b-9 concentration difference) remained unchanged during treatment. Prefilter concentrations of cytokines were constant during HF and cytokines were not detected in the ultrafiltrate. Prefilter C3a and C5a concentrations declined significantly during HF in patients and in volunteers. However, only C3a appeared in the ultrafiltrate. Ultrafiltrate from patients enhanced endotoxin-induced TNF-α release of Mo (table 1). On the other hand, PHA-induced

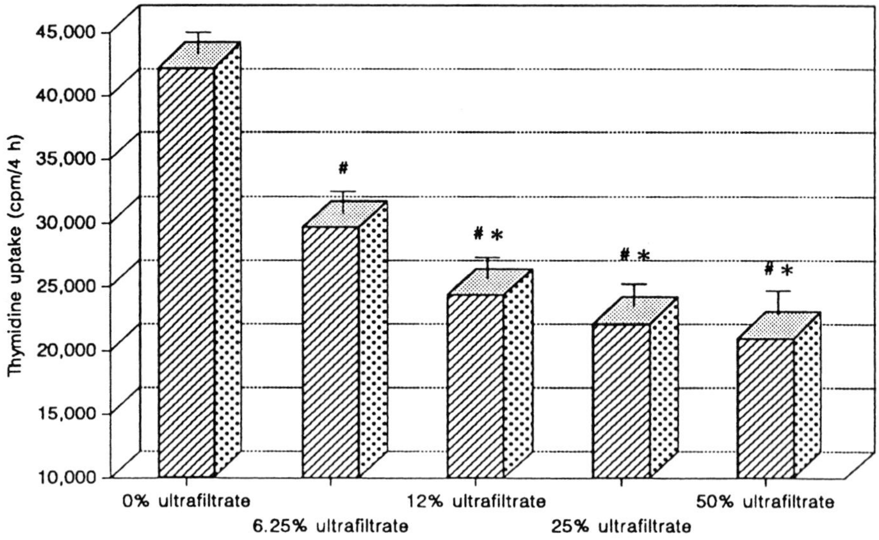

Fig. 1. Suppression of PHA-induced peripheral blood mononuclear cell (PBMC) proliferation by ultrafiltrate from septic patients in vitro. PBMC were incubated with increasing final concentrations of ultrafiltrate and proliferation was quantified by measuring tritiated thymidine uptake during 4 h. $^{\#}$ $p < 0.001$ vs. 0% ultrafiltrate; * $p < 0.001$ vs. 6.25% ultrafiltrate. cpm = Counts per minute.

IL-2 and IL-6 production of Ly was dramatically suppressed by ultrafiltrate from patients (table 1). Also, PHA-induced PBMC proliferation was suppressed by ultrafiltrate from patients compared to NaCl. The suppressive effect of ultrafiltrate from septic patients depended on the concentration of ultrafiltrate in the preparation (fig. 1). The more diluted, the less pronounced the suppression of PBMC proliferation was. Ultrafiltrate from volunteers did not cause any significant changes in the cell culture systems.

Conclusions

Systemic inflammatory response syndrome and MODS are thought to be a result of a self-enhancing generalized reaction, mediated by several cascade systems [6]. In the present study, we tested the hypothesis whether HF could eliminate some of these mediators, thereby rendering HF useful as an anti-inflammatory strategy in critical care medicine.

When HF is tested as a tool to eliminate mediators of MODS, it has to be excluded that the procedure itself causes substantial mediator activation by bioincompatibility of the HF membrane. According to recent studies, bioincompatibility is reflected by a rise in pre- and postfilter concentrations of C5b-9 [7]. Blood-membrane contact during HF did not lead to substantial activation of the complement system as indicated by our measurements.

In contrast, the anaphylatoxins C3a and C5a, but not the cytokines studied, were effectively eliminated by HF. We failed to demonstrate C5a in the ultrafiltrate samples despite the fact that the molecular weight of C3a and C5a is rather similar (C3a, 9 kD; C5a, 11 kD). Consequently, other mechanisms than filtration must have been involved in the fall of C5a concentrations during HF. It could be that C5a is rapidly adsorbed at the membrane or it is bound to cellular receptors of white blood cells as described previously [7].

Ultrafiltrate from patients with MODS contains mediators which altered healthy white cell function. The PBMC and the lymphocyte response to PHA in standard cell culture systems were significantly suppressed by ultrafiltrate from septic patients; on the other hand, ultrafiltrate from septic patients enhanced the endotoxin-triggered TNF synthesis. These changes of cell function are similar to changes observed in the cells from septic patients [5]. Consequently, HF eliminates compounds from the septic organism which convert healthy white blood cells into septic cells. This HF effect may be beneficial for patients' outcome.

In conclusion, HF represents a new modality of mediator removal and may be tested in clinical practice in situations calling for lowering anaphylatoxin concentrations and for an attenuation of exaggerated monocyte TNF production, and for stimulation of suppressed lymphocyte function.

References

1 Storck M, Hartl WH, Zimmerer E, Inthorn D: Comparison of pump-driven and spontaneous continuous hemofiltration in postoperative acute renal failure. Lancet 1991;337:452–455.

2 Grootendorst AF, van Bommel EFH: The role of hemofiltration in the critically ill intensive care unit patient: Present and future. Blood Purif 1993;11:209–223.

3 Bellomo R, Tipping P, Boyce N: Continuous hemofiltration with dialysis removes cytokines from the circulation of septic patients. Crit Care Med 1993;21:522–526.

4 Elebute EA, Stoner HB: The grading of sepsis. Br J Surg 1983;70:29–31.

5 Munoz C, Carlet J, Fitting C, Misset B, Bleriot J-P, Cavaillon J-M: Dysregulation of in vitro cytokine production by monocytes during sepsis. J Clin Invest 1991;88:1747–1754.

6 Bone RC: The pathogenesis of sepsis. Ann Intern Med 1991;115:457–469.

7 Deppisch R, Schmitt V, Bommer J, Hänsch GM, Ritz E, Rauterberg EW: Fluid phase generation of terminal complement complex as a novel index of bioincompatibility. Kidney Int 1990;37:696–706.

D. Inthorn, MD, Professor of Surgery and Intensive Care Medicine, Chirurgische Klinik, Klinikum Grosshadern, Marchioninistrasse 15, D–81377 München (Germany)

Sieberth HG, Stummvoll HK, Kierdorf H (eds): Continuous Extracorporeal Treatment in Multiple Organ Dysfunction Syndrome. Contrib Nephrol. Basel, Karger, 1995, vol 116, pp 80–85

Complement Fragments and Cytokines: Production and Removal as Consequences of Hemofiltration

Didier Journois

Department of Anesthesia and Intensive Care Medicine, Laennec Hospital, Paris, France

The clinical presentation of patients with multiorgan failure caused by septic conditions is very similar to that seen in patients with multiorgan failure after cardiopulmonary bypass (CPB) for cardiac surgery [1]. This similitude is denoted by the term of systemic inflammatory response syndrome (SIRS) which can lead to the development of adult respiratory distress syndrome [2]. The release of a large variety of inflammatory mediators may be the hallmark of this syndrome. There seems to be a synergy between these different mediators that may play a role individually, but may have a synergistic interaction and may be more detrimental in combination [3, 4]. Several works have pointed out that hemofiltration, as hemodialysis and CPB circuits, are able to induce the liberation of these mediators. CPB appears therefore as an interesting model of the SIRS pathogenesis [5]. Nevertheless, this deleterious effect seems to be widely counterbalanced in the specific case of hemofiltration by its ability to remove some of these mediators from plasma following their generation in sepsis or CPB. The raised questions are therefore to demonstrate the clinical usefulness of this removals and then to enhance the hemofiltration removal properties.

The Inflammatory Response to Extracorporeal Circuits

The human complement system includes numerous plasma components and regulatory proteins [6]. The contact of blood with a foreign surface triggers complement activation through either of two pathways resulting in the generation of C3 convertase. This enzyme acts on the third complement component resulting in the formation of C3a and C3b. This cleavage is amplified and C3b has the ability to transiently bind to exogenous surfaces. An accumulation of C3b molecules on the complement-activating surface occurs and results in the opsonization of the surface and to an enhanced reactivity with effector cells expressing C3 receptors. These receptors mediate secondary effector responses as elimination of opsonized targets or triggering of inflammatory reaction. The anaphylatoxins C3a and C5b are responsible for the generation of acute inflammatory mediators from leukocytes. A noticeable complement activation takes place in any extracorporeal circuit during hemodialysis, hemofiltration, or CPB, by increasing order of importance. This phenomenon may be reduced by using 'biocompatible' membranes made of polymethacrylate, polysulfone or polyacrylonitrile [7]. Unfortunately these latter materials cannot be used with the gas exchange membranes dedicated to CPB and the complement activation remains about 3-fold higher than with most recent hemodialysis membranes. Nevertheless, rates of complement activation during CPB are reduced by hypothermia. However, the CPB rewarming period is characterized by a huge release of C3a and C5b comparatively to the one observed in severe septic states, except that this peak is usually transient [8, 9].

Cellular Response and Cytokine Release

A transient but severe neutropenia occurs within the first minutes of extracorporeal circulation. The intensity of this neutropenia is highly correlated with complement activation and is therefore limited when hemodialysis or hemofiltration is performed with non-complement-activating membranes [10]. The complement fragments C5a and C5adesArg have been shown to be responsible for pulmonary sequestration of neutrophils, explaining the observed peripheral leukopenia [5]. This leukopenia involves *up-regulation* of the CD11/CD18 adhesive protein expression on the granulocyte membrane following binding of C5adesArg to specific receptors of these cells [11]. Stimulation of phagocytes with opsonized cells by C3 fragments or by activated complement results in the release of oxygen metabolites mediating cytotoxic effects by generation of superoxide anion (O_2^-), hydrogen peroxide (H_2O_2) or hydroxyl radical (OH·) [6]. Similarly, numerous receptors to complement fragments on neutrophil or monocyte membranes are involved in *cytokines* release.

Interleukin (IL)-1 is involved in the generation of the inflammation reaction and in immunoregulation. IL-1β remains intracellular and is released after stimulation by circulating monocytes and macrophages. The IL-1β release is well documented [12, 13], and follows by 20 h exposure to CPB in adult cardiac surgery patients [8]. IL-1 exerts its effects on lymphoid cells and induces fever, sepsis-like hemodynamic shock, loss of body weight, prostaglandin release and expression of procoagulant and adhesion properties by endothelial cells [6]. The biological effects of IL-1 are regulated by almost one IL-1 receptor antagonist (IL-1Ra) which is a potent inhibitor released by monocytes. IL-1 shares numerous systemic effects with tumor necrosis factor (TNF). These two cytokines are highly synergistic. Several reports have demonstrated that TNF, IL-6 and IL-8 release are stimulated by CPB in adults [8] and even more frequently in children [14, 15]. The correlation between IL-8 release and the length of CPB has been demonstrated [14]. These cytokines share numerous biological activities. A set of autoamplification pathways exists. A soluble form of TNF receptor acts as an antagonist. IL-8 is suspected to be a trigger of neutrophil-induced endothelial injury and therefore responsible for some of the sepsis- or CPB-induced pulmonary adverse effects [14].

Inflammatory Mediator Removal by Hemofiltration

Complement Fragment Removal

Andreasson et al. [9] have recently confirmed the activation of the complement cascade during pediatric CPB. Moreover, these authors performed ultrafiltration following CPB and noticed high concentrations of C3a and C5a in the ultrafiltrate. Our recent comparative study suggested that ultrafiltration performed during pediatric CPB is actually able to reduce C3a and C5a plasma levels [16]. Moreover, this removal seemed to be linearly correlated with the amount of withdrawn ultrafiltrate [16]. This phenomenon confirms the existence of a convective removal of complement fractions by ultrafiltration. Additional complement activation produced by the ultrafiltration circuit may occur during the use of ultrafiltration membranes [7]. This has led to the preference of high biocompatibility membranes for renal support in critically ill patients. Effectively, polyacrylonitrile hemofilters are known to induce less complement activation than polyamide or polysulfone hemofilters [7]. Regarding CPB, the amount of complement activation induced by the ultrafiltration device is probably negligible comparative to those induced by the gas-exchange membrane [5]. Nevertheless, one of the most interesting aspects of complement fragment removal by ultrafiltration during CPB may be the limitation of the delayed effects that these components have on cytokine release and therefore on the amplification of the inflammatory cascade [8]. Unfortunately, the importance of complement fraction removal could be reduced in situations of complement activation that are not due to transient contact of blood with an exogenous material as in sepsis states.

Cytokine Removal

The elimination of TNF-α and IL-1β by continuous ultrafiltration [17] and the beneficial effects of continuous ultrafiltration on cardiac and pulmonary functions during sepsis states have already been reported [18–20]. Moreover, the vol-

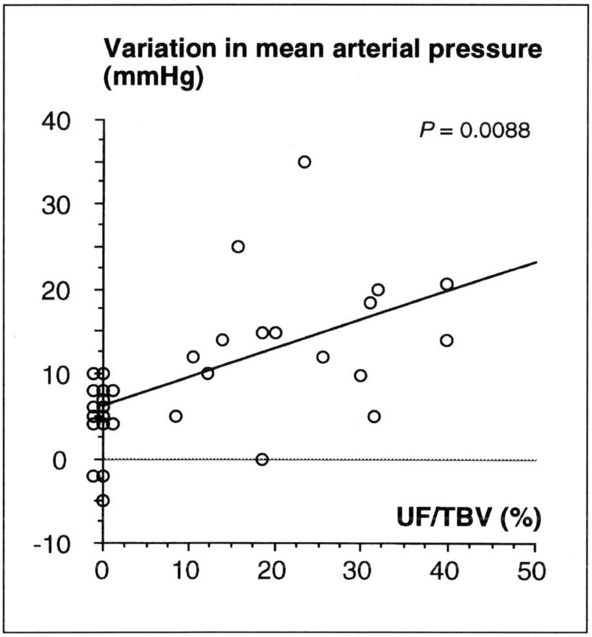

Fig. 1. Correlation between the amount of ultrafiltration expressed by the ratio of ultrafiltrated volume by the estimated total blood volume of circuit + patient (UF/TBV) and the improvement of mean arterial pressure between the beginning and the end of the rewarming period of CPB in two groups of children (the nonultrafiltered group is on the Y-axis). Spearman rank correlation test.

ume of daily ultrafiltrate could play an important role in organ function improvement in sepsis conditions [21]. Our recent study confirms the results from Millar et al. [15], showing that ultrafiltration is able to reduce plasma concentrations of IL-6 and TNF-α [22]. However, our results showed that IL-8 release is present during CPB rewarming but that IL-8 is poorly eliminated by ultrafiltration. Even if clinical variables like mean arterial pressure were significantly improved in children during hemofiltration, the responsibility of cytokine or complement removal for these effects was not established (fig. 1).

Cytokines, removed from plasma by ultrafiltration, are not necessarily only removed by convection. Barrera et al. [23] showed that incubation of polyacrylonitrile membrane fragments with radiolabeled IL-1β or TNF yielded a significant binding of both cytokines to the membrane. Removal was most marked in the first minutes, suggesting saturation of the membrane [23]. One of our studies showed that cytokine removals were not correlated with the amount of convective transport whereas it is the case for complement fragment removal. The potential rapid saturation of the cytokine binding process to hemofiltration membranes could limit its usefulness in clinical settings. Moreover, even if common sense argues for it, the usefulness of cytokine removal already has to be demonstrated ....

References

1 Casey LC: Role of cytokines in the pathogenesis of cardiopulmonary-induced multisystem organ failure. Ann Thorac Surg 1993;56:S92–S96.
2 Bone RC, Balk RA, Cerra FB, Dellinger RP, Fein AM, Knaus WA, Schein RM, Sibbald WJ: Definitions for sepsis and organ failure and guidelines for the use of innovative therapies in sepsis. The ACCP/SCCM Consensus Conference Committee. American College of Chest Physicians/Society of Critical Care Medicine. Chest 1992;101:1644–1655.
3 Tredget EE, Yu YM, Zhong S, Burini R, Okusawa S, Gelfand JA, Dinarello CA, Young VR, Burke JF: Role of interleukin-1 and tumor necrosis factor on energy metabolism in rabbits. Am J Physiol 1988;255:E760–E768.
4 Rothstein JL, Lint TF, Schreiber H: Tumor necrosis factor/cachectin. Induction of hemorrhagic necrosis in normal tissue requires the fifth component of complement. J Exp Med 1988;168:2007–2021.
5 Hammerschmidt DE, Weaver LJ, Hudson LD, Craddock PR, Jacob HS: Association of complement activation and elevated plasma-C5a with adult respiratory distress syndrome. Pathophysiological relevance and possible prognostic value. Lancet 1980;i:947–949.
6 Jahns G, Haeffner-Cavaillon N, Nydegger UE, Kazatchkine MD: Complement activation and cytokine production as consequences of immunological biocompatibility of extracorporeal circuits. Clin Mater 1993;14:303–336.
7 Mulvihill J, Cazenave JP, Mazzucotelli JP, Crost T, Collier C, Renaux JL, Pusineri C: Minimodule dialyser for quantitative ex vivo evaluation of membrane haemocompatibility in humans: Comparison of acrylonitrile copolymer, cuprophan and polysulphone hollow fibres. Biomaterials 1992;13: 527–536.
8 Haeffner-Cavaillon N, Roussellier N, Ponzio O, Carreno MP, Laude M, Carpentier A, Kazatchkine MD: Induction of interleukin-1 production in patients undergoing cardiopulmonary bypass. J Thorac Cardiovasc Surg 1989;98:1100–1106.
9 Andreasson S, Göthberg S, Berggren H, Bengtsson A, Eriksson E, Risberg B: Hemofiltration modifies complement activation after extracorporeal circulation in infants. Ann Thorac Surg 1993;56: 1515–1517.
10 Jacob HS, Craddock PR, Hammerschmidt DE, Moldow CF: Complement-induced granulocyte aggregation: An unsuspected mechanism of disease. N Engl J Med 1980;302:789–794.
11 Arnaout MA, Hakim RM, Todd R, Dana N, Colten HR: Increased expression of an adhesion-promoting surface glycoprotein in the granulocytopenia of hemodialysis. N Engl J Med 1985;312: 457–462.
12 Haeffner-Cavaillon N, Cavaillon JM, Ciancioni C, Bacle F, Delons S, Kazatchkine MD: In vivo induction of interleukin-1 during hemodialysis. Kidney Int 1989;35:1212–1218.
13 Laude-Sharp M, Caroff M, Simard L, Pusineri C, Kazatchkine MD, Haeffner-Cavaillon N: Induction of IL-1 during hemodialysis: Transmembrane passage of intact endotoxins. Kidney Int 1990; 38:1089–1094.
14 Finn A, Naik S, Klein N, Levinsky RJ, Strobel S, Elliott M: Interleukin-8 release and neutrophil degranulation after pediatric cardiopulmonary bypass. J Thorac Cardiovasc Surg 1993;105:234–241.
15 Millar AB, Armstrong L, van der Linden J, Moat N, Ekroth R, Westwick J, Scallan M, Lincoln C: Cytokine production and hemofiltration in children undergoing cardiopulmonary bypass. Ann Thorac Surg 1993;56:1499–1502.

16 Journois D, Vaccaroni L, Rambur R, Lagarde M, Keita M, Pouard P: Removal of complement fragments by ultrafiltration during pediatric cardiopulmonary bypass; in Sieberth HG, Stummvoll HK, Kierdorf H (eds): Continuous Extracorporeal Treatment in Multiple Organ Dysfunction Syndrome. Contrib Nephrol. Basel, Karger, 1995, vol 116, pp 179–182.
17 Bellomo R, Tipping P, Boyce N: Continuous veno-venous hemofiltration with dialysis removes cytokines from the circulation of septic patients. Crit Care Med 1993;21:522–526.
18 Gomez A, Wang R, Unruh H, Light RB, Bose D, Chau T, Correa E, Mink S: Hemofiltration reverses left ventricular dysfunction during sepsis in dogs. Anesthesiology 1990;73:671–685.
19 Stein B, Pfenninger E, Grunert A, Schmitz JE, Hudde M: Influence of continuous haemofiltration on haemodynamics and central blood volume in experimental endotoxic shock. Intensive Care Med 1990;16:494–499.
20 Stein B, Pfenninger E, Grunert A, Schmitz JE, Deller A, Kocher F: THe consequences of continuous haemofiltration on lung mechanics and extravascular lung water in a porcine endotoxic shock model. Intensive Care Med 1991;17:293–298.
21 Journois D, Chanu D, Safran D: Pump-driven haemofiltration. Lancet 1991;337:985.
22 Journois D, Pouard P, Rolland B, Lagarde M: Ultrafiltration allows to reduce cytokine plasma concentrations during pediatric cardiopulmonary bypass; in Sieberth HG, Stummvoll HK, Kierdorf H (eds): Continuous Extracorporeal Treatment in Multiple Organ Dysfunction Syndrome. Contrib Nephrol. Basel, Karger, 1995, vol 116, pp 86–88.
23 Barrera P, Janssen EM, Demacker PN, Wetzels JF, van der Meer JW: Removal of interleukin-1-beta and tumor necrosis factor from human plasma by in vitro dialysis with polyacrylonitrile membranes. Lymphokine Cytokine Res 1992;11:99–104.

Dr. D. Journois, Hôpital Laennec, 42, rue de Sèvres, F–75007 Paris (France)

Sieberth HG, Stummvoll HK, Kierdorf H (eds): Continuous Extracorporeal Treatment in Multiple Organ Dysfunction Syndrome. Contrib Nephrol. Basel, Karger, 1995, vol 116, pp 86–88

Ultrafiltration Allows to Reduce Cytokine Plasma Concentrations during Pediatric Cardiopulmonary Bypass

Didier Journois, Philippe Pouard, Bénédicte Rolland, Michel Lagarde

Department of Anesthesia and Intensive Care Medicine, Laennec Hospital, Paris, France

Ultrafiltration during cardiopulmonary bypass (CPB) in children has been reported to be an effective therapy to reduce the amount of accumulated tissue water [1]. Hemodynamic improvement [1] and improved ventricular contractility have been previously reported [2]. These studies suggest that the beneficial effects of ultrafiltration depend not only on the removal of water associated with CPB hemodilution but also on enhanced clearance of mediators of inflammatory reaction. This study was intended to determine the effects of ultrafiltration performed during pediatric CPB rewarming on plasma cytokines.

Material and Methods

Following ethical committee approval, 32 children undergoing surgical correction of tetralogy of Fallot were randomly assigned to be ultrafiltered during CPB rewarming or a control group. Ultrafiltration was performed with a polysulfone hemofilter (HFT02®; Sorin Lab.) during the whole rewarming period. Arterial blood sampling was obtained prior to rewarming (and ultrafiltration) (T_1) and at rewarming completion (T_2). Blood samples for tumor necrosis factor (TNF-α) and interleukin assays were withdrawn at T_1 and T_2. Sample supernatants were removed and allotted to polypropylene tubes of 300 μl following centrifugation (6 min, 3,000 *g*, 4 °C) then stored at –70 °C and assayed within 3 months. Enzyme-linked immunoassays (Elisa) for IL-6, IL-8 and TNF-α were performed using Quantikine® kits produced by Research & Diagnostic Systems (Minneapolis, Minn., USA). Their sensitivities were less than 5, 5 and 10 pg/ml respectively. IL-1β was assayed by Elisa (Cistron Biotechnology, Pine Brook, N.J., USA) (sensitivity < 5 pg/ml). Data were analyzed between groups using the Mann-Whitney nonparametric method for unpaired data. Statistical significance was chosen at $p < 0.05$.

Table 1. Evolution between T_2 and T_3 (rewarming ± hemofiltration period) of plasma concentrations of TNF, IL-1β, IL-6 and IL-8 in the two groups

	T_2 mean ± SD	T_3 mean ± SD	Between T_2 and T_3[a], p	Between groups, p
TNF, ng/ml				
UF	3.128±0.88	2.786±1.221	0.003	*
Control	2.758±0.474	3.283±0.707		*
IL-6, ng/ml				
UF	1.461±0.481	0.784±0.429	0.001	0.003
Control	1.593±0.549	1.596±0.54		*
IL-8, ng/ml				
UF	1.362±0.681	1.367±0.685	*	*
Control	0.969±0.412	0.971±0.424		*
IL-1, ng/ml				
UF	1.133±0.473	1.047±0.421	*	*
Control	1.3±0.141	1.325±0.035		*

[a] Mann-Whitney nonparametric method for unpaired data.

Results

TNF, IL-6 and IL-8 were detected in all patients from both groups. The plasma concentrations of IL-6, IL-8 and TNF-α were 1.5 ± 0.5, 1.1 ± 0.6 and 2.9 ± 0.7 ng/ml before rewarming respectively. A significant difference was noticed between groups with respect to TNF-α and IL-6 levels during rewarming, whereas no difference was observed in IL-8 levels (table 1). The levels of TNF-α, IL-6 and IL-8 were correlated with the ultrafiltration duration ($p = 0.01$). IL-1β activity was detected in only 3 patients undergoing ultrafiltration and in 2 who did not undergo ultrafiltration. These patients were seriously hypoxemic before the procedure (oxygen saturation (median [min–max]): 68 [54–73]%), and were among those presenting the higher levels of C5a at T_2 (241 [227–278] ng/ml vs. 189 ± 28 ng/ml in the rest of the group). IL-1β levels appeared to decrease in the ultrafiltration group but the small number in this data set did not permit any statistical testing.

Discussion

The numerous similarities between post-CPB morbidity and sepsis syndrome have led several groups to investigate cytokine release during and after CPB [3]. Several reports have demonstrated that TNF, IL-6 and IL-8 release are stimulated by CPB [4, 5]. Our study confirms the results of Millar et al. [4] showing that ultrafiltration, performed during the late phase of rewarming, is able to reduce the concentrations of IL-6 and TNF-α [4]. These substances removed from plasma by ultrafiltration are not necessarily only removed by convection. Barrera et al. [6] showed that incubation of polyacrylonitrile membrane with radiolabeled IL-1β or TNF yielded a significant and saturable binding of both cytokines to the membrane. IL-1β is an endogenous pyrogen and a key mediator of inflammation generated by monocytes. IL-1β remains intracellular and is released after stimulation by circulating monocytes. The IL-1β release is well documented in patients undergoing sequential hemodialysis, and follows exposure to CPB in adult cardiac surgery patients [7]. This delay could be explained either by the fact that IL-1β release is triggered by complement activation or by a suppressive effect of hypothermia [7]. IL-8 is suspected to be a trigger of neutrophil-induced endothelial injury and therefore responsible for some of the postoperative CPB adverse effects [2]. The correlation between IL-8 release and the length of CPB has been demonstrated [2]. Our results confirm that IL-8 release is present during rewarming and that IL-8 is poorly eliminated by ultrafiltration with the membrane used in this study. The ability of ultrafiltration to remove some major mediators of the inflammatory response could reduce the morbidity associated with CPB, and needs to be investigated further.

References

1 Naik SK, Knight A, Elliott M: A prospective randomized study of a modified technique of ultrafiltration during pediatric open-heart surgery. Circulation 1991;84(suppl III):422–431.

2 Elliott MJ: Ultrafiltration and modified ultrafiltration in pediatric open heart operations. Ann Thorac Surg 1993;56:1518–1522.

3 Casey LC: Role of cytokines in the pathogenesis of cardiopulmonary-induced multisystem organ failure. Ann Thorac Surg 1993;56:S92–S96.

4 Millar AB, Armstrong L, van der Linden J, Moat N, Ekroth R, Westwick J, Scallan M, Lincoln C: Cytokine production and hemofiltration in children undergoing cardiopulmonary bypass. Ann Thorac Surg 1993;56:1499–1502.

5 Haeffner-Cavaillon N, Cavaillon JM, Ciancioni C, Bacle F, Delons S, Kazatchkine MD: In vivo induction of interleukin-1 during hemodialysis. Kidney Int 1989;35:1212–1218.

6 Barrera P, Janssen EM, Demacker PN, Wetzels JF, van der Meer JW: Removal of interleukin-1-beta and tumor necrosis factor from human plasma by in vitro dialysis with polyacrylonitrile membranes. Lymphokine Cytokine Res 1992;11:99–104.

7 Haeffner-Cavaillon N, Roussellier N, Ponzio O, Carreno MP, Laude M, Carpentier A, Kazatchkine MD: Induction of interleukin-1 production in patients undergoing cardiopulmonary bypass. J Thorac Cardiovasc Surg 1989;98:1100–1106.

Dr. D. Journois, Hôpital Laennec, 42, rue de Sèvres, F–75007 Paris (France)

Sieberth HG, Stummvoll HK, Kierdorf H (eds): Continuous Extracorporeal Treatment in Multiple Organ Dysfunction Syndrome. Contrib Nephrol. Basel, Karger, 1995, vol 116, pp 89–98

Effect of Continuous Hemodiafiltration on IL-6, TNF-α, C3a, and TCC in Patients with SIRS/Septic Shock Using Two Different Membranes

Norbert Braun[a], *Silke Rosenfeld*[a], *Martina Giolai*[e], *Wolfgang Banzhaf*[b], *Reinhold Fretschner*[b], *Heiner Warth*[b], *Christoph Weinstock*[c], *Reinhold Deppisch*[d], *Christiane M. Erley*[a], *Gerhard A. Müller*[a], *Teut Risler*[a]

Departments of
[a] Internal Medicine III,
[b] Anesthesia and
[c] Transfusion Medicine, University of Tübingen, FRG;
[d] Gambro, FRG, and
[e] Institute of Clinical Microbiology and Immunology, St. Gallen, Switzerland

Septic shock, especially due to gram-negative infection, continues to be associated with high rates of morbidity and mortality despite advances in intensive care technologies, reviewed by Bone [1]. Several trials have sought to evaluate anticytokine and anti-LPS strategies with conflicting results [2–6], although administration of anticytokine monoclonal antibodies or soluble cytokine receptors prior to challenge with LPS in animal models provide essentially complete protection against shock and death [7–10].

Previous studies have claimed that continuous hemofiltration may be of benefit in septic patients with multiorgan dysfunction [11–15]. It was suggested that hemodiafiltration can remove systemic cytokines from the circulation in septic patients [16]. But there is very little information available on the outcome and clinical parameters in these patients compared to patients without hemofiltration. We therefore undertook a randomized, controlled study to assess the effect of continuous hemodiafiltration in patients with septic shock. In this paper we report the outcomes, clinical sepsis scores, cytokine and complement levels related to hemodiafiltration or maximal conservative treatment.

Methods

Trial Design and Selection of Patients

Written consent was only obtained from patients who underwent elective surgery. In all other cases, inclusion into the study was in the responsibility of the registrar on duty. Patients between 18 and 79 years of age who were intubated and mechanically ventilated for at least 48 h were included into the study if the criteria for the systemic inflammatory response syndrome (SIRS) modified according to the ACCP/SCCM consensus conference [17] and Bone [18] were present. Patients with SIRS were eligible for randomization to either hemodiafiltration (CVVHD) or maximal conservative treatment if they fulfilled the criteria of septic shock with multiple systems organ failure. Randomization was withdrawn if the patient experienced deterioration of renal function or if there was striking clinical evidence that the patient might profit from CVVHD, i.e. for fluid balance, parenteral nutrition or therapy of congestive heart failure. Patients with a primary renal, immunological or hematological disorder were excluded from the trial. Primary end point was death due to septic shock. Survival was defined as discharge from the hospital. The original study protocol was approved by the institutional review board of the University of Tübingen.

Hemodiafiltration

CVVHD was carried out on the BSM 22 device (Hospal, Nürnberg, Germany) with the BSM 20-CA111 for the arterial line and the BSM-V132 for the venous line, respectively. A polyamide membrane, FH66D (Gambro, Hechingen, Germany), or an acrylonitrile membrane, MF60 (Hospal), was used. Both membranes did not differ with respect to surface area (0.6 m^2), sieving coefficients or mode of sterilization (ETO). Blood flow was adjusted between 100 and 120 ml/min. Substitution fluid consisted of a balanced electrolyte solution containing 160 ml of bicarbonate 8.4% and 3 mmol/l of lactate (Schiwa Dialoc 4500 ml Basislösung SH44-HEP; Schiwa, Glandorf, Germany). Potassium was added to a final concentration of 4.5 mmol/l except in the cases of electrolyte imbalance. CVVHD was stopped whenever a patient improved markedly, i.e. the criteria for SIRS as mentioned above did not further apply for at least 48 h, and the patient was not dependent on renal replacement therapy for renal failure.

Plasma Samples and Measurement of Endotoxin, Tumor Necrosis Factor-α (TNF-α), Interleukin (IL)-6, Complement Component (C3a) and Terminal Complement Complex (TCC)

Blood was drawn in EDTA collecting tubes at intervals of 12 h from every patient with SIRS, immediately centrifuged at the bedside, and plasma samples were stored in aliquots at −70 °C until assayed within 3 months. Endotoxin in plasma was determined by the Limulus amebocyte lysate (LAL) test as described elsewhere [19], optimized by performing an internal standardization in each sample to exclude false-positive results due to nonendotoxin specific reactions [20]. TNF-α and IL-6 were assayed with ELISA systems purchased from Amersham, UK. C3a was determined by microaffinity chromatography using the ABICAP human complement-peptide test C3a/C3a-desArg (Abion, Benediktbeuren, Germany). TCC was measured with a micro-ELISA using a monoclonal antibody against neoantigens on the TCC, and highly purified human TCC serving as standard [21]. All measurements were carried out twice and the mean was used for further calculations.

Statistical Analysis

Variances were compared by the F test. Fisher's exact two-tailed test was used to test nominal parameters for difference between hemofiltration and control group, i.e. multiple organ dysfunction scores. Age, initial Apache III, Elebute score and serum creatinine, maximal endotoxin concentration and multiorgan failure scores were tested for significant differences in both groups using Student's two-tailed t test with Satherwaith's approximation for nonnormally distributed values. Regression analysis was performed on log-transformed cytokine, C3a and TCC concentrations in individual patients. The group means of the slopes were used for multiple two-sided t tests. A Bonferroni post-hoc comparison was performed to determine differences in cytokine and complement slopes as well as for trends in clinical scores. Data obtained at day 0 and day 3 were also compared by multiple t tests with Bonferroni post-hoc comparison. Survival was demonstrated by Kaplan-Meier curves [22] and compared by a log-rank analysis [23].

Results

Thirty patients (8 females, 22 males) were eligible for randomization to either hemodiafiltration or control therapy. In 6 cases (20%) the randomization was withdrawn and a different mode of therapy was started because hemodiafiltration seemed to be necessary to correct temporarily fluid overload without renal failure (2 cases), to prevent further progression into uremic state (2 cases) or to eliminate blood urea in a catabolic situation (1 case). In 1 case, hemofiltration was not started due to the overall poor prognosis of the patient. In total, 15 patients were treated maximally conservatively and 15 patients received additional hemodiafiltration. Patients' characteristics are summarized in table 1. Both groups did not differ with respect to age, initial Elebute scores, maximal endotoxin concentration or initial serum creatinine. Patients in the hemodiafiltration group had significantly more affected organs, i.e. a higher multiorgan failure score, at the beginning of the study compared with the controls (4.3 ± 0.8 vs. 3.2 ± 1.2) and a higher Apache III score. Cardiac surgery predominated among the underlying disease categories in both groups. All patients had either a positive blood culture, a positive LAL test for endotoxin or both.

In the CVVHD group, 10 patients survived and left the hospital while only 9 patients survived in the control group. The median survival time among the nonsurvivors was 11 days for CVVHD patients and 9 days for controls (fig. 1).

During the first 3 days of hemodiafiltration the Apache III score (Δ_{0-3} = −3.40, p = 0.0043), the Elebute score (Δ_{0-3} = −2.93, p = 0.011), the concentration of IL-6 (Δ_{0-3} = −2.04, p = 0.05) and TCC (Δ_{0-3} = −2.82, p = 0.01) declined significantly in the CVVHD group but not in the control group. None of these parameters showed a significant decrease in the control group. For TNF-α and C3a, no significant reduction could be observed in either group.

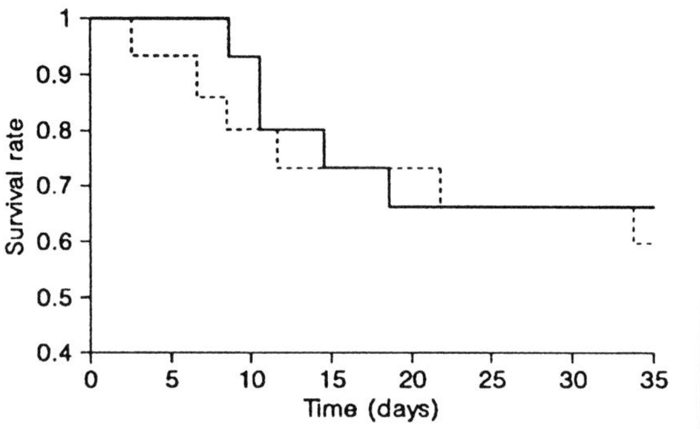

Fig. 1. Kaplan-Meier plot for survival rates of patients on CVVHD (——) or in the control group (– – –).

Table 1. Patients' characteristics at study entry

	CVVHD	Control	p value[1]
Male/female ratio	10/5	12/3	
Age, years	54.8 ± 14.6	51.3 ± 12.0	0.47
Initial Apache III score	92.5 ± 18.2	74.0 ± 11.4	<0.01
Initial Elebute score	9.9 ± 4.4	9.2 ± 3.8	0.62
Initial serum creatinine, mg/dl	1.4 ± 1.4	1.3 ± 0.8	0.9
Initial number of affected organs	4.3	3.2	<0.05[2]
Patients with creatinine > 1.5 mg/dl	3	4	1.0[2]
Maximal endotoxin concentration, pg/ml	958 ± 1,438	1,353 ± 2,702	0.62

[1] p values for the two-sided Student's t test.
[2] Fisher's exact test for nominal data.

In the whole course of treatment the differences among the groups only remained for the Apache III score. Elebute scores were not different any longer using the linear regression modelling of individual courses (table 2). Analyzing the subgroups of surviving patients in the CVVHD and control group, there still was a more rapid improvement in the CVVHD group ($slope_{Apache\ III}$: −2.8 ± 1.9) compared with the controls (−0.80 ± 0.93, p = 0.01). Measurements of IL-6 and TCC showed no significant differences in their slopes but TNF-α reduction was more evident in the CVVHD patients than in the controls ($slope_{ln\ (TNF\text{-}\alpha)}$: −0.2 vs. 0.24, p = 0.03). These differences also remained when data were stratified according to the initial multiorgan failure score.

Table 2. Mean trend slopes (±SD) for clinical scores, IL-6, TNF-α, activated C3a and TCC

	CVVHD	Control	p value[1]	MSD[2]
Apache III	−1.8±2.4	0.06±1.5	0.02	1.43*
Elebute	−0.14±0.35	−0.29±0.42	0.30	0.24
ln(IL-6)	−0.22±0.43	−0.13±0.21	0.47	0.25
ln(TNF-α)	−0.20±0.24	0.08±0.4	0.03	0.19*
ln(C3a)	−0.029±0.16	−0.024±0.1	0.92	0.094
ln(TCC)	−0.04±0.14	−0.01±0.07	0.49	0.088

For abbreviations, see text.

[1] p values denote significance levels obtained by Student's test for unpaired group means.

[2] Minimum significant difference (MSD) obtained by Bonferroni's post-hoc comparison of weighted data stratified by initial multiorgan dysfunction score, * $p < 0.05$.

Table 3. Mean trend slopes (±SD) of patients treated with either the FH66D (polyamide) or the MF60 (acrylonitrile) membrane

	FH66D	MF60	p value[1]
Apache III	−1.7±3.2	−1.3±1.7	0.76
Elebute	−0.31±0.58	−0.24±0.28	0.76
ln(IL-6)	−0.21±0.25	−0.1±0.09	0.29
ln(TNF-α)	−0.2±0.29	−0.08±0.08	0.35
ln(C3a)	−0.02±0.09	−0.02±0.04	0.92
ln(TCC)	−0.05±0.14	−0.03±0.06	0.83

For abbreviations, see text.

[1] p values denote significance levels obtained by Student's test for unpaired group means.

Among the patients in the CVVHD group, 8 patients received hemodiafiltration using the FH66D membrane and 7 patients were treated by the MF60 membrane. One patient on a FH60 membrane had to be treated temporarily with a AV600 membrane (Fresenius) and was excluded from further calculations. Although the Apache III score improved more clearly in patients on FH66D within the first 3 days (FH66D: −14.9; MF60: −2.1), there was no difference in the reduction rate of IL-6, TNF-α, C3a and TCC. Analysis of the complete individual courses showed no differences between the membranes (table 3).

Discussion

Multiple organ failure due to septic shock ranks among the most leading causes of death in intensive care units. Its mortality did not change during the last century [24]. The incidence of SIRS and its sequela, septic shock, is still rising because increasing numbers of patients with altered host defenses are admitted and susceptible to infections due to the widespread use of intravenous catheters and invasive cardiovascular monitoring [25].

The role of TNF-α and IL-6 as well as parts of the complement system, i.e. C3a, in the pathogenesis of septic shock is well established [26–30]. Bacterial cell wall enodotoxins are capable of stimulating the monocyte/macrophage system to secrete TNF-α which triggers further immune response, such as induction of TNF-α itself, IL-1 and IL-6. Enthusiastic reports on the reduction of mortality in patients with gram-negative sepsis receiving monoclonal antibodies towards endotoxins, i.e. HA-1A or E5 [4, 5], must be revised critically since a recently published, larger multicenter trial failed to confirm the previous results on HA-1A [6]. Another report on recombinant IL-1 receptor antagonist in patients with septicemia [2] stated that there was no difference in the outcome of patients treated either with IL-1 receptor antagonist or receiving placebo. Other studies on immunotherapy directed against TNF-α are also rather disappointing [31].

There exists substantial controversy about the selective suppression or removal of single mediators of SIRS/septic shock, because their natural and beneficial role in the host immune response is crucial [32]. Very often, the target of anticytokine strategies is present only for a very short time and on-line measurement of the target antigens is not available. Thus, the effect of intravenous anticytokine antibodies is limited by the unknown correct application time, the uncertainty of the duration of treatment and the probable inhibition of natural immune therapy mechanisms. A method which down-regulates the whole cytokine network for a certain period of time without affecting a specific mediator pathway could provide the intensivists with the needed time to look out for the cause of sepsis and appropriate treatment. This method should work in a continuous manner and must be operated in a simple and safe way. Experimental [33–35] and clinical studies [11, 12, 15, 36–39] encourage the use of prophylactic hemofiltration although definite proof of a benefit in patients with SIRS/septic shock has not yet been established.

We decided to investigate the benefit of the CVVHD as such a cytokine-reducing method in a relatively homogenous group of patients with SIRS/septic shock. Although CVVHD is less effective than pure hemofiltration in eliminating molecules with a molecular weight between 20 and 50 kD, this was the appropriate technique in our study design since it allowed us to continue the same renal replacement therapy even in the presence of developing renal failure without changing the protocol.

Even in the absence of positive blood cultures, endotoxin was involved in all our study patients. We did not find a difference in survival or mean survival time between patients under CVVHD or control therapy. This might be contributory to the small number of patients, since there was a significant difference in the initial Apache III score and the number of affected organs in the multiorgan dysfunction syndrome between patients on CVVHD and patients in the control group, indicating that CVVHD patients were more critically ill compared to their controls. This can be explained by the request of the institutional review board not to withhold a possible beneficial therapy to control patients which forced us to treat 5 patients with CVVHD despite randomization to the control group. Development of the clinical scores Apache III and Elebute showed a clear decline in both groups for patients who survived the study. The Apache III slopes of the patients in the CVVHD group declined more rapidly than in the control group. The Elebute score slopes might not differ because their increment is much rarely causing only small changes from day to day. So Elebute scores were not good parameters for day-to-day evaluation but showed clearly the septic state of the patients in the beginning.

Over the whole study period we found a significant difference between CVVHD and control treatment only regarding the TNF-α slope. TNF-α concentrations decreased more rapidly in the CVVHD group regardless of survival. The rapid reduction of the TNF-α concentration in patients on CVVHD can probably not be contributed to sole filtration as assumed by Bellomo et al. [16], but also to the adhesion of active TNF-α trimer to the membrane [40]. Thus a transient increase of TNF-α during the course of sepsis in the CVVHD group can easily be adsorbed by the filters while it can trigger a fatal immune response with further release of other relevant mediators in patients treated conservatively.

The absence of a substantial decrease of C3a, an indicator of sepsis and pulmonary dysfunction, in patients undergoing hemodialfiltration can readily be explained by the continuous activation of the alternative complement pathway due to the contact of blood with foreign material [41]. Since patients in the control group were less critically ill and their survival chance was higher compared to the patients in the CVVHD group, it has been expected that complement activation should come down soon in the natural course.

Interestingly, we did not find differences for IL-6 and TCC, both mediators of an advanced state of immune response to infection and autodysregulation. IL-6, a nonspecific cytokine in the cascade of TNF-α and many other factors, is decreased by effective conservative means in both groups. Unfortunately the concentrations of IL-6 show a wide range of distribution even in the individual patient within very few days. This reflects the release of IL-6 as an unspecific marker of many different immunological actions and counteractions. On the other hand, IL-6 is maximally increased in the beginning of multiorgan failure due to sepsis and seems to be harmful to the patient in this state [42, 43]. Analysis of IL-6 levels in the first 3 days after developing multiorgan failure shows a significant reduction in IL-6 levels only in the CVVHD patients. IL-6 is clearly filtered by polyamide and acrylonitrile membranes (data not shown). Similar results were reported by Gueugniaud et al. [44] for the elimination of IL-6 by CVVHD in severely burned patients.

For TCC, which cannot be filtered in contrast, we could not see any difference in slopes of TCC during the whole course of the study. TCC slopes tend to decrease more rapidly within the first 3 days in patients on CVVHD, reflecting a downregulation of the alternative complement activation in the more distant part of the complement system.

We also investigated the influence of either the polyamide or the acrylonitrile membrane on the mediator levels and slopes, respectively. Both membranes are known to be biocompatible and for both membranes it was shown that either convective or adsorptive removal of proinflammatory cytokines takes place [40]. There was no difference regarding the material of the filter despite the patients' groups not being homogenous. Elimination slopes were similar in both subgroups.

Clinical observations as well as several reports have suggested that in patients with SIRS/septic shock or multiple organ dysfunction the present clinical status of the patient may benefit from the early institution of a renal replacement therapy in order to remove cytotoxic substances, proinflammatory mediators or stabilize metabolic dysregulation. Our study confirm these observations. We furthermore could demonstrate that CVVHD has an important influence on IL-6 and TCC in the early state of septic shock and can permanently lower the TNF-α concentration. Although overall survival did not differ between both groups, we contribute this to the higher Apache III score and the higher number of failed organ systems in the CVVHD group. Thus, CVVHD provides doctors in intensive care units with an extracorporeal supportive technique for the additional treatment of septic-induced multiple organ failure patients which is safe and easy to handle. It may prolong the time between the onset of multiple organ failure and death in certain patients for specific measures against the cause of sepsis.

References

1 Bone RC: Gram-negative sepsis: A dilemma of modern medicine. Clin Microbiol Rev 1993;6:57–68.

2 Fisher CJ Jr, Dhainaut JF, Opal SM, Pribble JP, Nalk RA, Slotman GJ, Iberti TJ, Rackow EC, Shapiro MJ, Greenman RL, Reines HD, Shelly MP, Thompson BW, LaBrecque JF, Catalano MA, Knaus WA, Sadoff JC: Recombinant human interleukin-1 receptor antagonist in the treatment of patients with sepsis syndrome. JAMA 1994;271:1836–1843.

3 Fisher CJ Jr, Opal SM, Dhainaut JF, Stephens S, Zimmerman JL, Nightingale P, Harris SJ, Schein RM, Panacek EA, Vincent JL: Influence of an anti-tumor necrosis factor monoclonal antibody on cytokine levels in patients with sepsis. The CB0006 Sepsis Syndrome Study Group. Crit Care Med 1993;21:318–327.
4 Greenman RL, Schein RMH, Martin MA, Wenzel RP, MacIntyre NR, Emmanuel G, Chmel H, Kohler RB, McCarthy M, Plouffe J, Russel JA: A controlled clinical trail of E5 murine monoclonal IgM antibody to endotoxin in the treatment of gram-negative sepsis. JAMA 1991;266:1097–1102.
5 Ziegler EJ, Fisher CJ, Sprung CL, Staube RC, Sadoff JC, Foulke GE, Wortel CH, Fink MP, Dellinger RP, Teng NNH, Allen E, Berger HJ, Knatterud CL, LoBuglio AF, Smith CR: Treatment of gram-negative bacteremia and septic shock with HA-1A human monoclonal antibody against endotoxin. N Engl J Med 1991;324:429–436.
6 McCloskey RV, Straube RC, Sanders C, Smith SM, Smith CR: Treatment of septic shock with human monoclonal antibody HA-1A: A randomized, double-blind, placebo-controlled trial. Ann Intern Med 1994;121:1–5.
7 Ohlsson K, Bjork P, Bergenfeldt M, Hageman R, Thompson RC: Interleukin-1 receptor antagonist reduces mortality from endotoxin shock. Nature 1990;348:550–552.
8 Martin RA, Silva AT, Cohen J: Effect of anti-TNF-alpha treatment in an antibiotic treated murine model of shock due to *Streptococcus pyogenes.* FEMS Microbiol Lett 1993;110:175–178.
9 Bertini R, Delgado R, Faggioni R, Gascon MP, Ythier A, Ghezzi P: Urinary TNF-binding protein (TNF soluble receptor) protects mice against the lethal effect of TNF and endotoxic shock. Eur Cytokine Netw 1993;4:39–42.
10 Dinarello CA, Gelfand JA, Wolff SM: Anticytokine strategies in the treatment of the systemic inflammatory response syndrome. JAMA 1993;269:1829–1835.
11 Barzilay E, Kessler D, Berlot G, Gullo A, Geber D, Zeev IB: Use of extracorporeal supportive techniques as additional treatment for septic-induced multiple organ failure patients. Crit Care Med 1989;17:634–637.
12 Inthorn D, Storck M, Hartl WH, Zimmerer E: Verbesserung der Überlebensrate des postoperativen Nierenversagens durch hochvolumige Hämofiltration. Zentralbl Chir 1992;116:961–968.
13 Storck M, Hartl WH, Zimmerer E, Inthorn D: Comparison of pump-driven and spontaneous continuous haemofiltration in postoperative acute renal failure. Lancet 1991;337:452–455.
14 Storck M, Hartl WH, Inthorn D: Septic shock and multiple organ failure: Treatment with haemofiltration? Intensive Care Med 1991;17:372–373.
15 Blinzler L, Hausser J, Bodeker H, Zaune U, Martin E, Gebhardt C: Conservative treatment of severe necrotizing pancreatitis using early continuous venovenous hemofiltration. Contrib Nephrol. Basel, Karger, 1991, vol 93, pp 234–236.
16 Bellomo R, Tipping P, Boyce N: Continuous veno-venous hemofiltration with dialysis removes cytokines from the circulation of septic patients. Crit Care Med 1993;21:522–526.
17 Members of the American College of Chest Physicians/Society of Critical Care Medicine Consensus Conference Committee: Definitions for sepsis and organ failure and guidelines for the use of innovative therapies in sepsis. Crit Care Med 1992;20:864–874.
18 Bone RC: Toward an epidemiology and natural history of systemic inflammatory response syndrome. JAMA 1992;268:3452–3455.
19 Roth RI, Levin FC: Optimization of detection of bacterial endotoxin in plasma with the Limulus test. J Lab Clin Med 1990;116:153–161.
20 Urbaschek R, Becker KP: Endotoxinnachweis im Plasma: Spezifität und Aussagekraft für Entwicklung und Prognose einer Sepsis. Infusionsther Transfusionsmed 1993;20(suppl 1):16–20.
21 Deppisch R, Schmitt V, Bommer J, Hansch GM, Ritz E, Rauterberg EW: Fluid phase generation of terminal complement complex as a novel index of bioincompatibility. Kidney Int 1990;37:696–706.
22 Kaplan EL, Meier P: Nonparametric estimation from incomplete observations. J Am Stat Assoc 1958;53:457–481.
23 Mantel N: Evaluation of survival data and two new rank order statistics arising in its consideration. Cancer Chemother Rep 1966;50:163–170.

24 Werdan K: Neue Aspekte der Sepsis-Behandlung – Additive Therapiemassnahmen mit Antikörpern und Antagonisten gegen Bakterientoxine und Sepsismediatoren sowie mit Immunglobulinen. Intensivmedizin 1993;30:201–217.
25 Dahmash NS, Chowdhury MNH, Fayed DF: Septic shock in critically ill patients: Aetiology, management and outcome. J Infect Dis 1993;26:159–170.
26 Dinarello CA: The proinflammatory cytokines interleukin-1 and tumor necrosis factor and treatment of the septic shock syndrome. J Infect Dis 1991;163:1177–1184.
27 Endo S, Inada K, Inoue Y, Kuwata Y, Suzuki M, Yamashita H, Hoshi S, Yoshida M: Two types of septic shock classified by the plasma levels of cytokines and endotoxin. Circ Shock 1992;38:264–274.
28 Bellomo R: The cytokine network in the critically ill. Anaesth Intensive Care 1992;20:288–302.
29 Dofferhoff AS, de Jong HJ, Bom VJ, van der Meer J, Limburg PC, de Vries-Hospers HG, Marrink J, Mulder PO, Weits J: Complement activation and the production of inflammatory mediators during the treatment of severe sepsis in humans. Scand J Infect Dis 1992;24:197–204.
30 Zentella A, Manogue K, Cerami A: The role of cachectin/TNF and other cytokines in sepsis. Prog Clin Biol Res 1991;367:9–24.
31 Cohen J, Exley AR: Treatment of septic shock with antibodies to tumour necrosis factor. Schweiz Med Wochenschr 1993;123:492–496.
32 Hamilton G, Hofbauer S, Hamilton B: Endotoxin, TNF-alpha, interleukin-6 and parameters of the cellular immune system in patients with intraabdominal sepsis. Scand J Infect Dis 1992;24:361–368.
33 Lee PA, Matson JR, Pryor RW, Hinshaw LB: Continuous arteriovenous hemofiltration therapy for *Staphylococcus aureus*-induced septicemia in immature swine. Crit Care Med 1993;21:914–924.
34 Stein B, Pfenninger E, Grunert A, Deller A: Influence of continuous hemofiltration on hemodynamics and pulmonary function in porcine endotoxic shock. Contrib Nephrol. Basel, Karger, 1991, vol 93, pp 105–109.
35 Gomez A, Wang R, Unruh H, Light RB, Bose D, Chau T, Correa E, Mink S: Hemofiltration reverses left ventricular dysfunction during sepsis in dogs. Anesthesiology 1990,73:671–685.
36 Marzi I, Ziegenfuß Th, Rose St, Bauer M, Riegel W: Einfluss der kontinuierlichen veno-venösem Hämofiltration auf die posttraumatische Entzündungsreaktion; in Riegel W (ed): Kontinuierliche Blutreinigungsverfahren in der Intensivmedizin. Lengerich, Pabst, 1994, pp 52–67.
37 Cosentino F, Paganini E, Lockrem J, Stoller J, Wiedemann H: Continuous arteriovenous hemofiltration in the adult respiratory distress syndrome. A randomized trial. Contrib Nephrol. Basel, Karger, 1991, vol 93, pp 94–97.
38 Rau HC, Staubach KH, Hohlbach C, Klingler W: The continuous arterio-venous hemofiltration in shock. Prog Clin Biol Res 1987;236B:241–271S.
39 Coraim FJ, Coraim HP, Ebermann R, Stellwag FM: Acute respiratory failure after cardiac surgery: Clinical experience with the application of continuous arteriovenous hemofiltration. Crit Care Med 1986;14:714–718.
40 Goldfarb S, Golper TA: Pro-inflammatory cytokines and hemofiltration membranes. J Am Soc Nephrol 1994;5:228–232.
41 Johnson RJ: Complement activation during extracorporeal therapy: Biochemistry, cell biology and clinical relevance. Nephrol Dial Transplant 1994;9(suppl 2):36–45.
42 Calandra T: High circulating levels of interleukin-6 in patients with septic shock: Evolution during sepsis, prognostic value, and interplay with other cytokines. Am J Med 1991;91:23–38.
43 Friedland JS, Suputtamongkol Y, Remick DG, Chaowagul W, Strieter RM, Kunkel SL, White NJ, Griffin GE: Prolonged elevation of interleukin-8 and interleukin-6 concentrations in plasma and of leukocyte interleukin-8 mRNA levels during septicemic and localized *Pseudomonas pseudomallei* infection. Infect Immun 1992;60:2402–2408.
44 Gueugniaud PY, Bertin-Maghit M, Jolly MO, Hirschauer C, Bouchard C, Petit P: Rôle de l'interleukine-6 dans la phase oedémateuse du brûlé grave. Presse Méd 1993;22:735.

Dr. med. Norbert Braun, Medizinische Klinik und Poliklinik, Abteilung III,
Universitätsklinik Tübingen, Otfried-Müller-Strasse 10, D-72076 Tübingen (Germany)

Sieberth HG, Stummvoll HK, Kierdorf H (eds): Continuous Extracorporeal Treatment in Multiple Organ Dysfunction Syndrome. Contrib Nephrol. Basel, Karger, 1995, vol 116, pp 99–103

The Influence of Continuous Hemofiltration on Cytokine Elimination and the Cardiovascular Stability in the Early Phase of Sepsis

A. Sander, W. Armbruster, B. Sander, Th. Philipp, Ch. Schäfer, N. Thürauf, R. Lange

Departments of Anesthesiology, Nephrology and General Surgery, University Medical School, Essen, Germany

The typical hemodynamic situation of patients with sepsis can be characterized as a decrease of systemic vascular resistance in combination with an increase of cardiac output in case of adequate fluid resuscitation [1]. The underlying generalized vasodilation results from an augmented production of nitric oxide by vascular endothelium cells [2]. A distinct relaxation of the vascular smooth muscles appears, which leads to a severe decrease in mean arterial blood pressure. To maintain a sufficient organ perfusion, it is often necessary to use vasopressor agents.

It is well known that in the course of sepsis, proinflammatory cytokines like TNF-α are involved in the generation of hypotension by inducing the nitric oxide production [3]. IL-6 is involved in regulating the inflammatory response. High circulating levels of IL-6 in the bloodstream are thought to be a marker for the severity of disease [4]. Furthermore, it could be shown that cytokines are eliminated from the bloodstream of septic patients during extracorporeal renal replacement therapy [5]. Beyond this there is evidence that extracorporeal support is of benefit for the hemodynamic stabilization in septic shock [6]. The studies mentioned above were experimental or dealt with patients who had already developed organ dysfunction such as acute renal failure. It is obvious that organ insufficiencies are partly generated by peripheral vascular disturbances [7]. The influence of continuous venovenous hemofiltration on the cardiovascular stability was examined as well as the development of plasma levels and the elimination rate of TNF-α and IL-6. It was aimed at counteracting the development of hemodynamic instability at the very beginning of the septic course in a prospective, controlled, randomized study. Hence, treatment started in the early phase of septic shock irrespective of the presence of acute renal failure.

Methods

After approval by the ethics committee of the University of Essen, 16 postoperative patients after major abdominal surgery were randomly assigned either to a control or a therapy group. Entrance criteria were those of the ACCP/SCCM consensus conference [8] and both a decrease in systemic vascular resistance and an increase in cardiac output. The patients serving as controls received conventional therapy corresponding to standardized ICU protocols. The patients of the therapy group additionally underwent continuous veno-venous hemofiltration for at least 48 h. For vascular access, a double-lumen central venous catheter was inserted either to a jugular or a subclavian vein. Hemofiltration was performed using a roller pump (BSM 22; Hospal, Lyon, France) and a filter with the AN 69 membrane (Multiflow 60®; Hospal). A blood flow of 150 ml/min led to a filtration rate of about 1,000 ml/h. The ultrafiltrate was replaced quantitatively by a bicarbonate-buffered crystalloid solution. Patients of the therapy group as well as the substitution fluid were warmed to avoid the influence of lowered body temperature on the hemodynamic parameters. Systemic anticoagulation with heparin (approximately 1,000 U/h) was controlled using the activated clotting time. All patients were equipped for invasive monitoring of hemodynamic variables using an arterial and a balloon-tipped, flow-directed pulmonary artery catheter (93A-431H-7,5F; Baxter-Edwards Critical Care, Irvine, Calif., USA). They received 2 μg/kg/min dopamine to support the splanchnic perfusion during mechanical ventilation. As soon as mean arterial blood pressure was <60 mm Hg (despite adequate fluid resuscitation) epinephrine was used. In case of a severe vasodilation, norepinephrine was used additionally to guarantee sufficient organ perfusion. Hemodynamic values were measured at the beginning and 12 and 48 h later. At the same time, blood samples (10 ml EDTA blood) were collected and after cold centrifugation stored at –70 °C. TNF-α and IL-6 were measured using an ELISA assay (Biosource Int., Camarillo, Calif., USA). Statistical analysis was performed by polynomial regression and Student's t test for paired and impaired data, respectively. A p value <0.05 was accepted to be significant.

Results

The severity of disease estimated from the initial Apache II values [9] was similar. The mean score was 14.0 ± 4.6 for the control group and 14.9 ± 4.8 for the filtration group (means ± SD). Plasma levels of TNF-α and IL-6 showed no significant differences between the two groups (fig. 1). In 5 patients, levels of TNF-α and IL-6 were examined in the filtrate. The amount of cytokines eliminated by the hemofiltration was calculated from the filtrate levels and the filtered volume. During the observed 48 h the estimated amount of TNF-α was 1.7 ± 0.4 μg, the amount of IL-6 came to 14.6 ± 8.7 μg (means ± SD). In spite of a

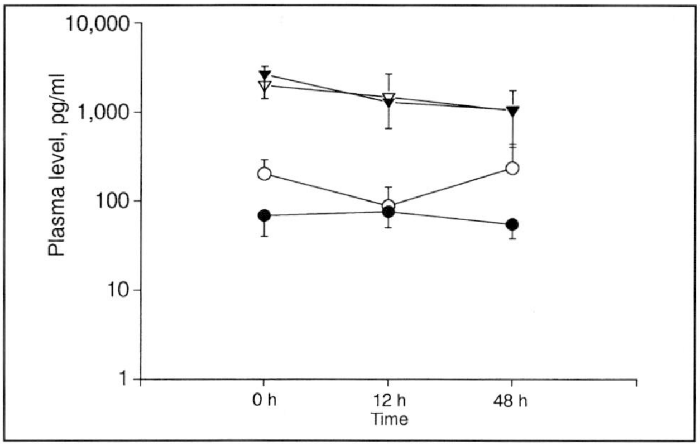

Fig. 1. Plasma levels of TNF-α and IL-6. ○ = TNF-α control group; ● = TNF-α filtration group: ▽ = IL-6 control group; ▼ = IL-6 filtration group (means ± SD).

marked elimination of cytokines, plasma levels did not decline in the filtration group.

Filling pressures (CVP/PCWP) 12–15 mm Hg) were kept constant after initial optimization in all patients to exclude the influence of fluid shifting on the circulatory regulation (fig. 2). The essential observation concerning the hemodynamic parameters was a significant increase of the systemic vascular resistance index in the filtration group during the 48 h of treament (fig. 3). While both groups started at about 1,000 $dyn \cdot s \cdot cm^{-5} \cdot m^{-2}$, vascular resistance index of the filtration group rose up to 1,400 $dyn \cdot s \cdot cm^{-5} \cdot m^{-2}$ ($p < 0.05$). No significant differences could be found in body temperature and the administration of catecholamines. In particular, there was no increased use of vasopressor agents in the filtration group at 48 h.

Discussion

Discussing the elimination rate under hemofiltration, it is important to reflect on the pharmacokinetics of the substances in question. The half-life time of cytokines is about 15 min [10]. This period of time seems to be rather short so that plasma levels possibly cannot be influenced in a considerable range by elimination with a hemofilter. The spontaneous decay might exceed the amount of cytokines eliminated by the filtration. Apart from cytokines found in the plasma, there are other compartments which cannot be detected using the ELISA assay. Cytokines are bound to the receptor site or found in a polymeric state of aggrega-

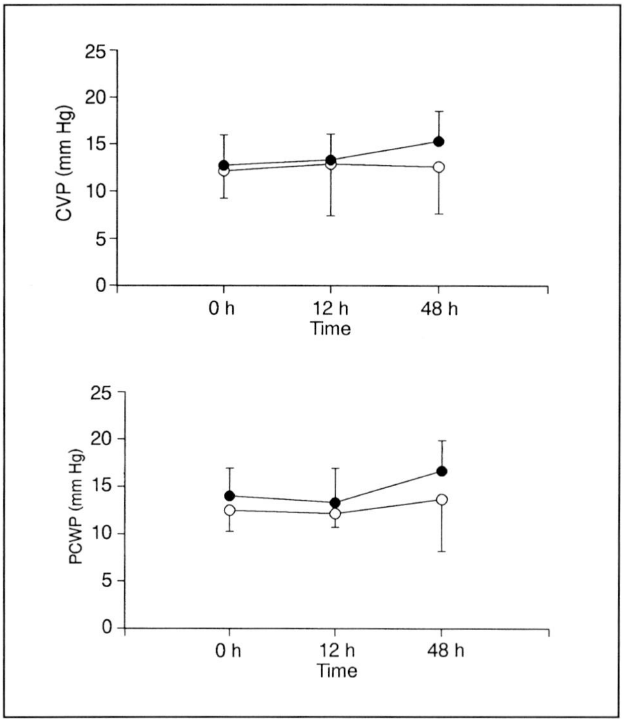

Fig. 2. Filling pressures (central venous pressure [CVP], pulmonary capillary wedge pressure [PCWP]). ○ = Control group (n = 8); ● = filtration group (n = 8) (means ± SD).

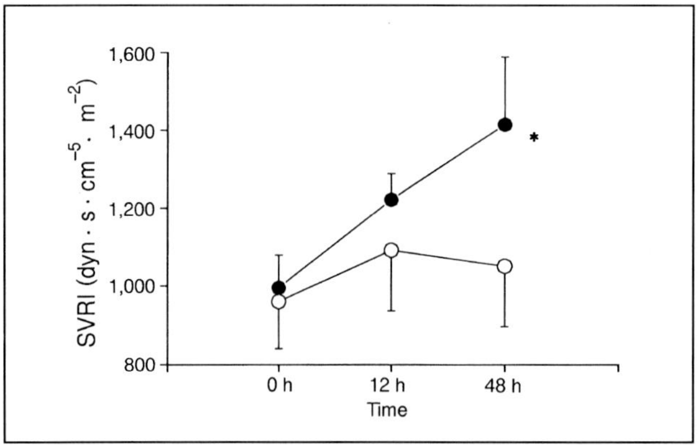

Fig. 3. Systemic vascular resistance index (SVRI). ○ = Control group; ● = filtration group (means ± SD). * $p < 0.05$ (polynomial regression).

tion [11]. The elimination of free substances might lead to a shifting from the polymeric to the detectable condition. Extracorporeal systems are able to induce production of cytokines and other hemodynamically relevant substances [12]. Therefore, biocompatible membranes are commonly used for extracorporeal support. Concerning the large amount of eliminated cytokines in this study, this factor is unlikely to play a major role.

The significant increase of the systemic vascular resistance index observed in the hemofiltration group cannot be attributed to changes in electrolytic levels, fluid shiftings, changes in blood temperature or different administration of catecholamines. Neither can the development of the plasma levels of TNF-α and IL-6 be correlated with these changes in the hemodynamic state. Whereas the underlying cause remains unclear, hemofiltration contributes to the stabilization of the cardiovascular system in the early phase of sepsis. There is a rationale for hemofiltration irrespective of renal function under the conditions of controlled studies to elucidate the underlying principles.

References

1 Parrillo J: The cardiovascular pathophysiology of sepsis. Annu Rev Med 1989;40:469–485.

2 Palmer RMJ: The discovery of nitric oxide in the vessel wall. A unifying concept in the pathogenesis of sepsis. Arch Surg 1993;128:396–401.

3 Kilbourn RG, Griffith OW: Overproduction of nitric oxide in cytokine mediated and septic shock. J Natl Cancer Inst 1992;84:827–831.

4 Damas P, Ledoux D, Nys M, Vrindts Y, de Groote D, Franchimont P, Lamy M: Cytokine serum level during severe sepsis in human. IL-6 as a marker of severity. Ann Surg 1992;215:356–362.

5 Bellomo R, Tipping P, Boice N: Continuous veno-venous hemofiltration with dialysis removes cytokines from the circulation of septic patients. Crit Care Med 1993;21:522–526.

6 Grootendorst AF, van Bommel EFH, van der Hoven B, van Leengoed LAMG, van Osta ALM: High volume hemofiltration improves right ventricular function in endotoxin-induced shock in the pig. Intensive Care Med 1992;18235–240.

7 Gutierrez G, Brown SD: Response of the macrocirculation, in Schlag G, Redl H (eds). Pathophysiology of Shock, Sepsis, and Organ failure. Berlin, Springer, 1993, pp 215–229.

8 ACCP/SCCM Consensus Conference Committe: in Bone RC, Balk RA, Cerra FA, Dellinger RP, Fein AM, Knaus WA, Schein RMH, Sibbald WJ (eds): Definitions for Sepsis and Organ Failure and Guidelines for the Use of Innovative Therapies in Sepsis. Chest 1992;101:1644–1655.

9 Knaus WA, Draper EA, Wagner DP, Zimmermann JE: Apache II, A severity of disease classification system. Crit Care Med 1985;13:818–829.

10 Fong Y, Moldawer LL, Shires T, Lowry S: The biologic characteristics of cytokines and their implication in surgical injury. Surg Gynecol Obstet 1990;170:363–378.

11 Wingfield P, Pain RH, Craig S: Tumour necrosis factor is a compact trimer. FEBS Lett 1987;211: 179–184.

12 Lonnemann G, van der Meer JWM, Cannon JG, Dinarello CA, Koch KM, Grannoleras C, Deschodt G, Shaldon S: Induction of tumor necrosis factor during extracorporeal blood purification. N Engl J Med 1987;317:963–964.

Dr. A. Sander, Institute of Anesthesiology, University of Essen, Hufelandstrasse 55, D-45147 Essen (Germany)

Sieberth HG, Stummvoll HK, Kierdorf H (eds): Continuous Extracorporeal Treatment in Multiple Organ Dysfunction Syndrome. Contrib Nephrol. Basel, Karger, 1995, vol 116, pp 104–107

Continuous Hemofiltration in Phase of Recovery from Septic Shock

E.M. Redl-Wenzl, P. Sporn

Department of Anesthesia and Intensive Care, Krankenanstalt Rudolfstiftung, Vienna, Austria

Capillary leakage in septic states is not only a problem of pulmonary microvasculature, but also of a vast majority of the whole body capillary bed [1]. The first and most urgent step in circulatory support is to normalize intravascular volume [2]. For this purpose, massive volume loading becomes necessary, as a large proportion of fluid administered escapes into the interstitial space.

During recovery from sepsis and re-establishment of capillary integrity, sequestrated fluid shifts back from interstitial into the intravascular space. In this stage, high negative fluid balances become mandatory to avoid cardiac congestion and extravascular pulmonary fluid accumulation. In case of renal failure, continuous renal support is of extreme importance. To elucidate this problem, a retrospective analysis of patients suffering from oliguric acute renal failure (ARF) recovering from sepsis was performed between January 1990 through March 1994.

Patients and Methods

Table 1 gives a survey of basic data. For pump-driven (NFG 05; Dialysetechnik, Karlsruhe, Germany) hemofiltration (HF 80; Fresenius) a double-lumen catheter was inserted in the internal jugular vein or femoral vein. Blood flow was regulated at flow rates between 100 and 160 ml/min, standard heparin was used to prevent clotting of the filter. Fluid replacement was performed with a specially designed electrolyte solution (HF21 and HF 02; Fresenius). Fluid balance was calculated every hour and steered by careful evaluation of pulmonary arterial occlusion pressure (PAOP), cardiac index (CI) and mean arterial pressure (MAP). Additionally, lung/chest X-ray was evaluated every 12 h for diagnosis of extravascular lung water accumulation. The main parameter for fluid withdrawal was serum colloid osmotic pressure (COP) to notice decrease due to backshift.

Table 1. Patients' characteristics

Patient	Age	Sex	Cause of septic shock	ICU stay days
1	49	F	Cholangitis	47
2	29	F	Gangrenous small bowel, peritonitis	78
3	31	F	Ovarian abscess, peritonitis	61
4	51	M	Superinfected necrotizing pancreatitis	91
5	53	M	Perforated small bowel, peritonitis	20
6	45	M	Anastomoses leakage, peritonitis	39
7	44	M	Superinfected intra-abdominal hematoma	54
8	57	M	Cholangitis	10
9	41	M	Meningitis	32
10	55	F	Meningococcal septicemia	27

Table 2. Fluid balances, hemodynamic parameters and COPs under continuous hemofiltration during recovery from sepsis

Day of recovery	Fluid balance ml/24 h	CI $l/min \cdot m^2$	PAOP mm Hg	MAP mm Hg	COP mm Hg
1	$-3,307 \pm 1,910$	4.89 ± 0.44	13.1 ± 1.9	84.8 ± 8.5	15.6 ± 0.8
2	$-5,410 \pm 3,084$	4.98 ± 0.5	12.4 ± 2.0	86.3 ± 8.4	16.2 ± 1.4
3	$-3,190 \pm 3,059$	4.79 ± 0.5	10.8 ± 1.8	85.9 ± 7.0	17.0 ± 1.7
4	$-3,980 \pm 4,010$	4.96 ± 0.5	10.7 ± 2.1	87.2 ± 7.7	17.3 ± 2.0
5	$-3,477 \pm 3,345$	4.98 ± 0.5	10.6 ± 1.8	86.4 ± 6.2	17.7 ± 1.9
6	$-2,566 \pm 2,436$	4.86 ± 0.4	10.4 ± 2.0	87.5 ± 6.5	18.5 ± 1.8
Mean 15 days (range 8–29)	cumulative balance –35 liters (range –9 to –80 liters)				

Values are means ± SD. For abbreviations, see text.

Results

Table 2 gives a survey of hemodynamic and renal data, serum COPs and fluid balances during the first days of recovery. All patients survived and could be discharged from ICU with stable renal function as shown in table 3.

Table 3. Renal function parameters and COP on day of discharge from ICU

Fluid balance ml/24 h	COP mm Hg	BUN mg/dl	Scr mg/dl	Ccr ml/min
315±353	22.4±1.3	44.4±24	1.9±1.2	41±26

BUN = Blood urea nitrogen; Scr = serum creatinine; Ccr = creatinine clearance. Values are means ± SD.

Table 4. Signs of backshift

PAOP ↑	COP ↓	GFR ↑
CI ↑		UF ↑

GFR = Glomerular filtration rate; UF = urinary flow.

Discussion

The extent of fluid sequestration from the intravascular into the interstitial space was underestimated during the past decades of intensive care therapy as interest was mainly focused on pulmonary capillary leakage and ARDS. This led to a rather restrictive fluid policy which counteracted optimized cardiac performance, oxygen transport and consumption. Data documenting early generalized microvascular injury [3] led to a considerable generous fluid administration. High positive fluid balances are well tolerated despite sharp decreases in COP and COP-PAOP gradient under the prerequisite of carefully steering low cardiac filling pressures [4]. This change in management of acute septic stages leads to the problems of fluid backshift during recovery. This is no problem if the signs of backshift (table 4) are well interpreted and treated with negative fluid balances supported by low-dose dopamine and loop diuretics. In case of oliguric ARF, continuous renal treatment becomes essential. A cumulative negative fluid balance of 351 (range 9–801) within 15 days as shown in table 2 can hardly be achieved by intermittent fluid withdrawal [5]. Our data demonstrate that optimal cardiac performance assigned by cardiac indices of nearly 5 l/min could be maintained under continuous fluid withdrawal. This is doubtlessly the main advantage of CVVHF. Additional actions by elimination of cytokines, mediators or toxic metabolites remain still speculative [6].

References

1 Ellman H: Capillary permeability in septic patients. Crit Care Med 1984;12:629–633.

2 Shoemaker WC, Appel PL, Kram H, Bishop M, Abraham E: Sequence of physiologic patterns in surgical septic shock. Crit Care Med 1993;21:1876–1889.

3 Hersh M, et al: Histologic and ultrastructural changes in nonpulmonary organs during early sepsis. Surgery 1990;107:397–410.

4 Zadrobilek E, Hackl W, Sporn P, Steinbereithner K: Effect of large volume replacement with balanced electrolyte solutions on extravascular lung water in surgical patients with sepsis syndrome. Intensive Care Med 1989;15:505–510.

5 Sporn P: Hämofiltrations- und Dialysemethoden in der operativen Intensivmedizin. Chirurg 1992; 63:993–998.

6 Laggner AN, Druml W, Lenz K, Schneeweiss B, Grimm G: Influence of ultrafiltration/hemofiltration on extravascular lung water; in Sieberth HG, Mann H, Stummvoll HK (eds): Continuous Hemofiltration. Contrib Nephrol. Basel, Karger, 1991, vol 93, pp 65–70.

E.M. Redl-Wenzl, Department of Anesthesia and Intensive Care, Krankenanstalt Rudolfstiftung, Juchgasse 25, A–1030 Wien (Austria)

Sieberth HG, Stummvoll HK, Kierdorf H (eds): Continuous Extracorporeal Treatment in Multiple Organ Dysfunction Syndrome. Contrib Nephrol. Basel, Karger, 1995, vol 116, pp 108–111

Effects of Continuous Venovenous Hemofiltration on Pulmonary Function and Hemodynamics in Postoperative Septic Multiorgan Failure

I. Krüger[a], *C. Jacobi*[a], *P. Landwehr*[b]

[a] Department of Surgery (Head: Prof. *H. Pichlmaier)* and
[b] Institute of Radiological Diagnostics (Head: Prof. *K. Lackner)* University of Cologne, Germany

In septic multiorgan failure, acute renal failure with hyperhydration and uremia leads to additional damage to the shocked lung. A septic invasion of endotoxins with release of vasoactive mediators leads to destabilization of hemodynamics. The intention of the study is to show that use of continuous venovenous hemofiltration (CVVH) leads to improvement of pulmonary function and stabilization of hemodynamics by flushing out endotoxin mediators and regulation of fluid balance with decrease of urea and creatinine.

Materials and Methods

Nineteen patients in whom CVVH was used to treat septic multiorgan failure in the period from October 1991 to September 1993 participated in the present study. The basis for the definition of sepsis was the sepsis score according to Elebute and Stoner [1]. Sepsis was regarded to be proved at a score count of > 12. We defined the multiorgan failure according to the model of Frey et al. [2].

We chose a Shaldon central venous double-lumen catheter as vascular access for the CVVH. We administered heparin at a dosage of 10 IU/kg bw/h for anticoagulation; this was introduced into the system in front of the HF filter (FH 88 H; Gambro). The pump volume of the hemofiltration system (BMM 10.1; Gambro) was regulated to obtain a fluid turnover of 40–45 liters/day.

We determined the effectiveness of CVVH on the basis of the decrease in the substances urine and creatinine and the filtration which is possible without impairment of the circulation. We measured the concentration of the endotoxin mediator serotonin in the ultrafiltrate in the hyperdynamic phase of sepsis in 3 patients. With regard to pulmonary function and its development on therapy with CVVH, we investigated the requisite FiO_2 of the patients with constant values of the measured blood gases. The decrease in pulmonary edema was controlled by an X-ray check of the chest every 24 h which was evaluated by a radiologist on the basis of a score system. Depending on the degree of severity of the finding (0: not detectable; 1: moderately pronounced; 2: very pronounced), the following pathological changes were considered in this additive X-ray score: pleural effusion (both sides appraised separately), pulmonary edema, enlargement of the heart, signs of pulmonary venous stasis, dilatation of the mediastinal veins, pulmonary infiltrates. Stabilization of hemodynamics should be detected by means of reduction of catecholamines during CVVH. Complications which may be connected with the CVVH (vascular injury, thrombosis, embolism, hemorrhage, catheter sepsis, catheter thrombosis) were documented.

Results

The average age of the patients was 59 ± 11 years. The youngest patient was 45 years and the oldest patient 85 years old. Altogether, CVVH was carried out on 64 days (on average 6.4 ± 1.12 days/patient). With a blood flow of 125 ± 31 ml/min, the fluid turnover of 40–45 liters/day aimed for could be attained. Under the specified heparinization, there was no significant impairment of blood clotting. A decrease of urea and creatinine could be detected especially at the beginning of CVVH. The average urea concentration was 192.6 ± 79.08 mg/dl at the beginning of CVVH and could be lowered within 4 days to an average value of 120.63 ± 42.9 mg/dl. Before therapy with CVVH, the average creatinine concentration was 3.87 ± 1.88 mg/dl and could be lowered to a value of 1.79 ± 0.56 mg/dl. Compared to a positive fluid balance before CVVH caused by the acute renal failure, a negative balance averaging 4.3 liters could be attained in the first 6 days of CVVH. In 3 patients, the endotoxin mediator serotonin and its degradation product 5-hydroxyindoleacetic acid was found at a concentration of 0.2 mg/dl in the ultrafiltrate in the hyperdynamic phase of sepsis.

The average FiO_2 in artificial ventilation was 0.57 ± 0.25 before CVVH and could be lowered to 0.30 ± 0.11 with constant values for blood gases under CVVH. In the radiological follow-up examinations, a decrease of the X-ray score under CVVH was documented. A comparison of the results before the beginning of CVVH with those on the sixth day of CVVH shows a significant reduction of the mean X-ray score from $x = 4.7$ to $x = 2.6$ ($p < 0.02$).

Stabilization of hemodynamics could be detected by means of reduction of catecholamines during CVVH. Dopamine could be reduced from 9.5 ± 3.7 to a minimum of 3.9 ± 1.9 µg/kg bw/min at day 8 of hemofiltration, dobutamine was reduced from 8.7 ± 3.2 to 1.2 ± 1.1 µg/kg bw/min at day 8.

There was initial hypotension at the beginning of CVVH in 3 patients in whom an attempt had been made to establish as high a negative balance as possible because of a substantial pulmonary edema. This hypotension could be eliminated by administration of an appropriate colloidal volume. In 2 patients, there was a catheter thrombosis of the proximal lumen, so that the catheter had to be changed.

After an intervening improvement, the general condition of the patient deteriorated once more as a result of the underlying septic multiorgan failure, so that 13 of our 19 patients died.

Discussion

Septic multiorgan failure is frequently associated with an acute renal failure and constitutes a severe postoperative complication with a high lethality. Despite improvements in intensive care in the last three decades, rates of lethality of 70–100% are reported in the literature [3, 4]. CVVH constitutes an alternative to the usual methods of blood filtration or dialysis which should be taken into consideration. Barzilay et al. [5] reports a very much better rate of survival in therapy with CVVH in contrast to other filtration methods in septic patients. In CVVH, the filtration performance is independent of fluctuations in the patient's blood pressure as in hemodialysis and can be carried out even under labile circulatory conditions. The continuous pump-driven filtration enables the amount of ultrafiltrate to be precisely precalculated, so that an adaption of fluid and electrolyte balance to clinical needs becomes possible. In contrast to hemodialysis, continuous dehydration relieving the circulation is feasible.

A positive fluid balance before CVVH could be reduced in a satisfactory way without stress on the patient's circulation. The labile circulation of the patient in septic multiorgan failure is to be mentioned again in this connection: it not uncommonly leads to hypotension and deterioration of cardiovascular function in hemodialysis. Besides the establishment of a fluid balance, the CVVH enables attainment of an adequate excretion of the substances urea and creatinine, as has also been described by List et al. [6]. With the detection of 5-hydroxyindoleacetic acid in the ultrafiltrate of 3 patients in the hyperdynamic phase of sepsis, the elimination of endotoxin mediators by CVVH could be demonstrated. This enables avoidance of additional damage to the alveolar membrane from the toxins specified. Gotloib et al. [7] report on the reduction of pulmonary edema in patients with ARDS by ultrafiltration of vasoactive agents, which is consistent with our investigations. Stein et al. [8] and Cheadle et al. [9] corroborate these results in which an improvement of organ functions (especially the lungs) could be attained by the hemofiltration of endotoxins in experimental animals in which sepsis had been artificially induced.

The radiological findings confirm that CVVH is effective by reducing pulmonary edema and pleural effusions, as documented by the significant improvement of the X-ray score. The introduction of the X-ray score we have developed thus provides a simple means of checking the effectiveness of CVVH.

The continuous dehydration of the patient with clinically and radiologically demonstrable reduction of pulmonary stasis and pulmonary edema which relieves the circulation and the excretion of substances which are usually eliminated via the urine as well as endotoxin mediators led to a marked improvement of pulmonary function in our patients. The FiO_2 raised before CVVH could be lowered to normal values of about $FiO_2 = 0.3$ with constant values for blood gases. The reduction of the catecholamine requirement and the improvement of cardiac function must be mentioned.

As has been shown, the complications we observed mainly concerned the CVVH system itself and could therefore easily be remedied. The clinical experience, a low rate of complications and the simple handling of CVVH indicates that it is a good alternative to other methods of hemofiltration for use in septic multiorgan failure. The demonstrable improvement of organ functions under CVVH enables the root cause of septic multiorgan failure to be controlled, thus lowering the lethality of this severe condition.

References

1 Elebute EA, Stoner HB: The grading of sepsis. Br J Surg 1983;70:29–31.
2 Frey D, Fultan R, Palk H: Multiple system organ failure. Arch Surg 1980;115:136–138.
3 Cameron J: ARF in the ICU. Intensive Care Med 1986;12:64–66.
4 Lazarus JM: Acute renal failure. Intensive Care Med 1986;12:61–63.
5 Barzilay E, Kessler D, Berlot G, Gullo A, Geber D, Been Zeer I: Use of extracorporeal supportive techniques as additional treatment for septic-induced multiple organ failure patients. Crit Care Med 1989;17:634–637.
6 List WF, Kulier A, Kiesling A, Semu J: Hämofiltration beim akuten Nierenversagen; Erfahrungen einer operativen Intensivstation. Anaesthesist 1990;39:540–546.
7 Gotloib L, Barzilay E, Shustak A, Lev A: Sequential hemofiltration in nonoliguric high capillary permeability pulmonary edema of severe sepsis: Preliminary report. Crit Care Med 1984;12:997–1000.
8 Stein B, Pfenninger E, Grünert A, Schmitz JE, Deller A, Kocher F: The consequences of continuous haemofiltration on lung mechanics and extravascular lung water in a porcine endotoxic shock model. Intensive Care Med 1991;17:293–298.
9 Cheadle WG, Hanasawa K, Gallinaro RN, Nimmanwudipong T, Kodama M, Polk HC Jr: Endotoxin filtration and immune stimulation improve survival from gram-negative sepsis. Surgery 1991;110:785–791.

Dr. I. Krüger, Department of Surgery, University of Cologne, Joseph-Stelzmann-Strasse 9, D–50924 Cologne (Germany)

Sieberth HG, Stummvoll HK, Kierdorf H (eds): Continuous Extracorporeal Treatment in Multiple Organ Dysfunction Syndrome. Contrib Nephrol. Basel, Karger, 1995, vol 116, pp 112–114

Continuous Monitoring of Cardiac Output during Hemodialysis

O. Bastien, S. Filley, S. Paulus, V. Piriou, S. Estanove

Anesthesia and Intensive Care Department, Hôpital Cardiovasculaire et Pneumologique, Lyon, France

Hemodynamic instability is a current problem during hemodialysis. Many causations can be involved [1, 2], like dysautonomia, hypovolemia, decrease of ionized calcium, buffer toxicity. Whatever, cardiac output (CO) can decrease because of a myocardial depressor factor, or increase simultaneously with a drop of vascular tone.

The aim of the study is to re-evaluate the interest of monitoring the CO during conventional hemodialysis, compared to standard monitoring of arterial pressure (AP) and heart rate (HR).

Materials and Methods

Ten patients after cardiac surgery are studied. All of them had an acute renal failure caused by acute tubular necrosis and were scheduled for first-time hemodialysis without severe low CO. For this reason, inotropic support could be unchanged during the procedure. Cardiac monitoring included direct AP measurement by radial artery catheterization, and right ventricular catheterization by a Swan-Ganz catheter via the internal jugular vein. This flow-directed pulmonary artery catheter has been modified (Vigilance®, Baxter) by placing a blood-warming filament in the region of the catheter injectate port. With an on-off sequence of heating, it is able to reconstruct continuously a thermodilution curve. This method has been validated and compared to standard methods in animals [3] and man [4].

Hemodialysis was performed via a double lumen venous access by a generator (Monitral®, Hospal), using bicarbonate buffer, pediatric lines and polyacrylonitrile membrane (PAN 69). The blood pump was carefully controlled in order to increase pump flow from 50 ml/min every 10 min up to 200 ml/min. Fluid balance was controlled so that hourly negative balance was 100 ml. CO was continuously monitored. At 7 times, simultaneously AP, HR, right AP, and pulmonary artery occlusive pressure (PAOP) were measured. Thermodilution CO was controlled at the same time by 3–5 measures in a random order during the

Table 1. Hemodynamic parameters during hemodialysis

	T0	T1	T2	T3	T4	T5	T6
CO, l/min	5.2±2.2	5.4±2.6	5.2±2.3	5.2±2.2	4.9±1.8	4.5±1.9*	4.3±1.8*
MAP, mm Hg	85±12	78±14	79±12	84±14	84±12	81±9	81±9
HR/min	95±24	96±23	94±22	97±23	96±23	96±21	103±27
Right AP, mm Hg	8.7±2	7.3±2	9.0±3	7.0±3	7.4±2	6.7±2	6.7±2
PAOP, mm Hg	13.9±5	11.2±3*	11±4	9.8±4	10±4	10.5±5	8.7±4*

* $p < 0.05$.

respiratory cycle. Times of measurement were T0 at the beginning, T1 to T4 for each blood flow step, and T5 and T6 after 1 h of steady state. Statistics included ANOVA for repeated measurements and Newman-Keuls test for a significant level of 0.05.

Results

Continuous CO measurements are strictly correlated (v = 0.944–0.986) to standard thermodilution techniques. The Blank and Altman test bias has been previously published [3, 4]. This technique has only recently been confirmed to be possible during hemodialysis. CO decreased significantly by 17% (table 1), without significant change of mean arterial pressure (MAP), HR, right AP and PAOP.

Discussion

Continuous measurement of CO is relevant during venovenous hemodialysis, in spite of possible blood thermal change or turbulent blood flow. In a cardiac patient without clinical low CO, monitoring AP and HR seems to be unable to detect a significant decrease of CO. This change appears to be more frequent after 2 h of dialysis, in spite of a specially smooth increase of blood flow. The mechanisms could not be precisely explained by Swan-Ganz monitoring, but do not seem to be related only to a change of right or left ventricular preload. Ventricular dysfunction could therefore be suspected, as has been proposed [5]. In case of more aggressive therapy, increasing blood pump flow up to 200 ml/min in a few minutes, a more severe cardiac dysfunction remains to be confirmed. Nevertheless, in spite of the progressive protocol, the hemodynamic stability is not so good as the continuous hemofiltration technique [6, 7].

Conclusion

Continuous CO monitoring may be useful in cardiac patients under hemodialysis, specially if this technique is preferred to continuous hemofiltration.

References

1 Bradley JR, Evans D, Cowley A: Comparison of vascular tone during combined haemodialysis with ultrafiltration and during ultrafiltration followed by haemodialysis. BMJ 1990;300:1312.

2 Henrich WL: Hemodynamic instability during hemodialysis. Kidney Int 1986;30:605–612.

3 Yelderman M, Quinn MD, McKown RC, Eberhart RC, Dollar ML: Continuous thermodilution cardiac output measurement in sheep. J Thorac Cardiovasc Surg 1992;104:315–320.

4 Yelderman M, Ramsay MA, Quinn MD, Paulsen AW, McKown RC, Gillman PH: Continuous thermodilution cardiac output measurement in intensive care unit patients. J Cardiothorac Vasc Anesth 1992;6:270–274.

5 Nixon JV, Mitchell JM, McPhaul JJ, Henrich WL: Effect of hemodialysis on left ventricular function: Dissection of changes in filling volume and in contractile state. J Clin Invest 1983;71:377–384.

6 Bastien O, Saroul C, Hercule C, George M, Estanove S: Continuous veno-venous hemodialysis after cardiac surgery. Contrib Nephrol. Basel, Karger, 1991, vol 93 pp 76–78.

7 Gullo A, Romano E, Mocavero G, Kette F, Berlot G: Cardiovascular function during continuous arteriovenous hemofiltration in critically ill patients. Int Symp CAVH. Milan, Wichtig, 1986, pp 313–317.

Dr. O. Bastien, Anesthesia and Intensive Care Department, Hôpital Cardiologique, boulevard Pinel, F–69394 Lyon Cedex 03 (France)

Sieberth HG, Stummvoll HK, Kierdorf H (eds): Continuous Extracorporeal Treatment in Multiple Organ Dysfunction Syndrome. Contrib Nephrol. Basel, Karger, 1995, vol 116, pp 115–120

Continuous Extracorporeal Treatment in Patients with Normal Renal Function?

Ludwig Blinzler[a], *Hermann Bödeker*[b], *Christoph Gebhardt*[b], *Dieter Heuser*[a]

Clinics for
[a] Anesthesiology and Surgical Intensive Medicine and
[b] Abdominal, Thoracic, and Endocrine Surgery, City Hospital Nuremberg, Germany

Since 1977, continuous arteriovenous hemofiltration has been established as a simple and safe therapy for acute renal failure (ARF). In patients with sepsis and multiple organ failure (MOF), pump-driven continuous venovenous hemofiltration (CVVH) allows better effectiveness in catabolic states [15]. Experimental [14] and clinical [5, 8] reports about blood purification and positive experience with CVVH in the conservative treatment of severe necrotizing pancreatitis [1] suggest that hemofiltration in addition to critical care basic therapy could have a beneficial effect in MOF, which is still associated with a poor prognosis in spite of pharmacological and technical progress.

Patients

From 1989 to 1991, 78 consecutive patients aged between 18 and 79 years (mean 58.5) were treated with CVVH. MOF was caused by hemorrhagic-traumatic noxa in 21% and by septic-toxic event in 79% (table 1). All patients required mechanical ventilation ($FiO_2 > 0.5$). 86% showed hyperdynamic shock with need of invasive monitoring and application of vasoactive drugs. 85% had ARF with elevated creatinine (>2.0 mg/dl) and oliguria/anuria. Hepatic failure with coagulopathy and icterus was seen in 53%. Thrombocytopenia appeared in 79%. Medium Apache II score amounted to 29.8 points.

Methods

Hemofiltration in MOF was installed when critical care basic therapy including adequate respiratory and hemodynamic management, parenteral nutrition, antibiotic treatment, etc., failed to stabilize organ functions. CVVH was added when pulmonary gas

Table 1. Genesis of MOF following different diseases

MOF in infection	n	MOF without infection	n
Pneumonia	21	Necrotizing pancreatitis	11
Peritonitis	15	Thoracoabdominal aortic tubing	6
Sepsis	8	Ruptured abdominal aortic aneurysm	4
Infected necrosis in pancreatitis	3	Gastrointestinal bleeding	4
cholangitis	2	Polytrauma	2
urosepsis	2		

exchange had $AaDO_2 > 350$ mm Hg or other problems like hyperdynamic shock, ARF, and hepatic failure appeared. CVVH was stopped when signs of MOF returned to normal range. Hemofiltration was performed from 6 h to 51 days, on the average 11.9 days.

Vascular access was obtained using a double lumen catheter inserted percutaneously into one of the great veins. CVVH was done in postdilution mode by a roller pump (BSM 22, Hospal) with blood flow of 200 ml/min and ultrafiltration rate of 33 ml/min (50 liters/day) and by a polyamide hollow-fiber membrane (FH 66, Gambro). Anticoagulation was achieved by sodium heparin from 0 to 2,500 U/h. Hemofilters had to be changed after the medium was run for 24 h.

Results

During consequent application of CVVH, most of these patients showed an improvement in their clinical course (fig. 1). Pulmonary stabilization was seen in 71% (55/78) (fig. 2). In 68% (46/67) vasoactive drugs could be reduced or discontinued. Anuria or oliguria changed to polyuria in 66% (44/66). Recovering from hepatic malfunction was seen in 37% (15/41). Thrombocytopenia normalized in 54% (33/62).

Thirty-three patients (42%) survived and were discharged from hospital. Twenty-two of 45 nonsurvivors showed temporary improvement of their clinical course. Six of them suffering from incurable malignancy unknown when starting CVVH died because of the underlying diseases. Ten of the nonsurvivors died because of fatal MOF within 24 h after admission to the ICU. Patients (n = 46) with early application of CVVH (within 48 h after the onset of MOF) showed a better survival rate (61%) than those (n = 32) who were treated later (16%). Mediators of MOF, i.e. products of the complement cascade measured in blood and ultrafiltrate by ELISA, were partially removed by CVVH. A C3a sieving coefficient of 0.11 was found.

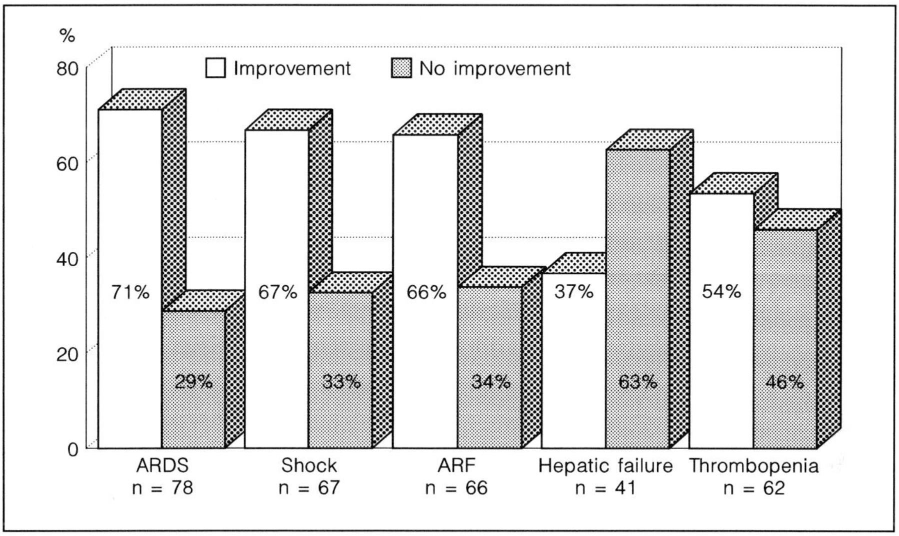

Fig. 1. Changes in organ functions during application of CVVH.

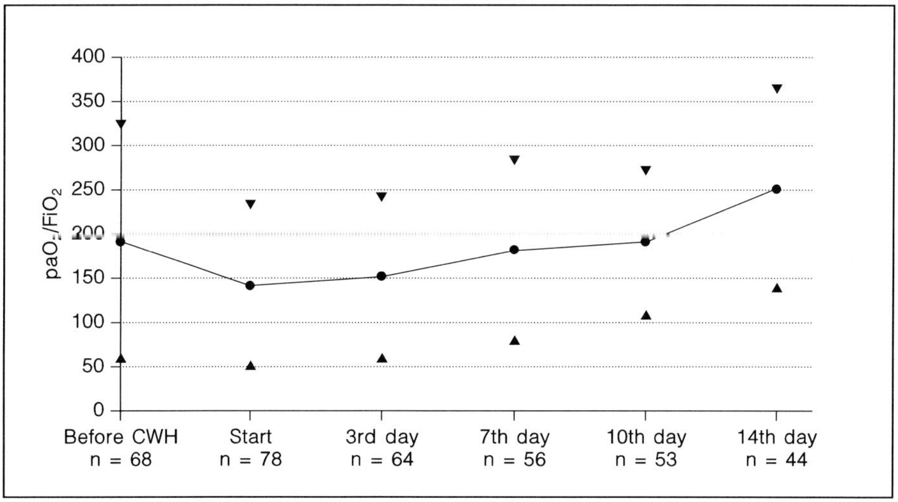

Fig. 2. Respiratory quotient during CVVH. Increasing pulmonary function with hemofiltration treatment. ● = Mean quotients.

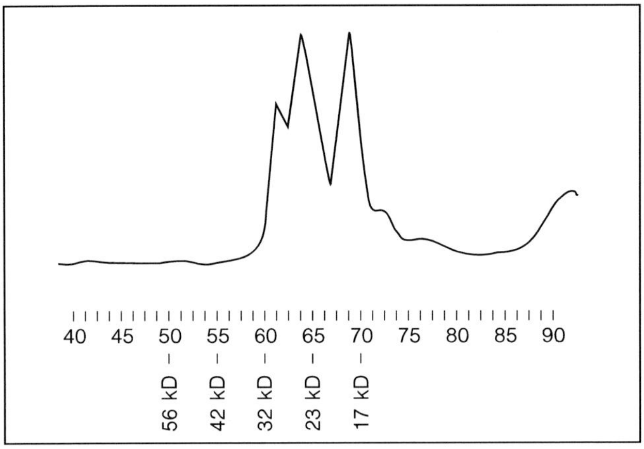

Fig. 3. HPLC analysis of UF in necrotizing pancreatitis, second day of CVVH. There are high peaks at 16–18 kD and others at 24–28 kD.

The daily testing of ultrafiltrate (UF) by HPLC demonstrated decreasing spikes of polypeptides during hemofiltration. Analyzing UF of patients with necrotizing pancreatitis showed high peaks at 16–18 kD and others at 24–28 kD (fig. 3). UF from septic/infected patients had high spikes at 10–12 and 28–34 kD and low levels at 17 kD (fig. 4).

Conclusions

MOF seems to be generated by cascade activation of immune competent cells and plasmatic mediators (e.g. bradykinin, eicosanoids, cytokines, anaphylatoxins) [2, 3]. Therapeutic approaches aim to inactivate or eliminate single substances by application of specific antibodies, receptor antagonism or suppression of the synthesis [2, 3]. CVVH with high-flux membranes [6] in combination with high-volume exchange allows elimination of many mediators with different molecular weights [4, 7, 9–13] by convective transport, and therefore may contribute to improving the prognosis of MOF. Other significant advantages of this technique such as adequate nutrition, optimized fluid balance and control of body temperature should not be neglected.

Filtration of substances is a physical process and depends on the hydrostatic and osmotic pressure, the membrane characteristics, the molecular size and loading, and the quotient of solute versus bound molecules. Polypeptides found in UF by HPLC analysis correspond to the free fraction of these proteins in plasma,

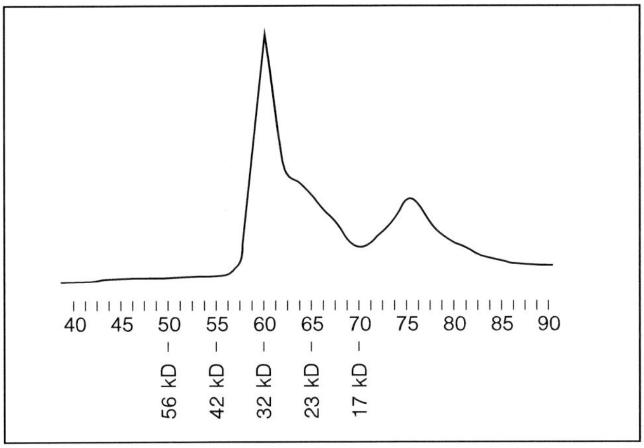

Fig. 4. HPLC analysis of ultrafiltrate in peritonitis, second day of CVVH. There are high spikes at 10–12 and 28–34 kD and low levels at 17 kD.

which is assumed to represent the bioactive substances in case of hormones and mediators.

UF of patients with systemic infections showed spikes which may include anaphylatoxins (C3a 9 kD, C5a 11 kD) and dimeric forms of cytokines like TNF (34 kD). In contrast to these findings, UF from patients with necrotizing pancreatitis without bacterial infection had high peaks at 16–18 kD, which may contain monomer cytokines like TNF and IL-1. The curves of UF HPLC analysis may possibly be helpful in distinguishing between the sterile and infected necrotizing pancreatitis. Furthermore, these findings support a hypothesis of different molecular biological mechanisms initiating and maintaining MOF in infectious and noninfectious diseases. Identification of the substances included in the different peaks and their importance for the generation of MOF are aims of future investigations.

References

1 Blinzler L, Hausser J, Bödeker H, Zaune U, Martin E, Gebhardt CH: Conservative treatment of severe necrotizing pancreatitis using early continuous venovenous hemofiltration; in Sieberth HG, Mann H, Stummvoll HK (eds): Continuous Hemofiltration. Contrib Nephrol. Basel, Karger, 1991, vol 93, pp 234–236.
2 Bone RC: The pathogenesis of sepsis. Ann Intern Med 1991;115:457–469.
3 Bone RC: A critical evaluation of new agents for the treatment of sepsis. JAMA 1991;266:1686–1701.

4 Coraim F, Wolner E: Management of cardiac surgery patients with continuous arteriovenous hemofiltration; in Sieberth HG, Mann H, Stummvoll HK (eds): Continuous hemofiltration. Contrib Nephrol. Basel, Karger, 1985, vol 33, pp 103–110.
5 Di Carolo JV, Dudley TE, Sherbotic JR, Kaplan BS, Costarino AT: Continuous arteriovenous hemofiltration/dialysis improves pulmonary gas exchange in children with multiple organ failure. Crit Care Med 1990;18:822–832.
6 Göhl H, Buck R, Strathmann H: Basic features of the polyamide membranes; in Shaldon S, Koch KM (eds): Polyamide – The evolution of a synthetic membrane for renal therapy. Contrib Nephrol. Basel, Karger, 1992, vol 96, pp 1–25.
7 Golper TA, Jenkins R, Wright M, Klein JB: Tumor necrosis factor and hemofiltration membranes (abstract). Intensivbehandlung 1990;15:119.
8 Gotloib L, Barzilay E, Shustak A, Lev A: Sequential hemofiltration in non-oliguric high capillary permeability pulmonary edema of severe sepsis: Preliminary report. Crit Care Med 1984;12:997–1003.
9 Hirasawa H, Sugai T, Ohtake Y, Oda S, Shiga H, Matsuda K, Kitamura N: Continuous hemofiltration and hemodiafiltration in the management of multiple organ failure; in Sieberth HG, Mann H, Stummvoll HK (eds): Continuous Hemofiltration. Contrib Nephrol. Basel, Karger, 1991, vol 93, pp 42–46.
10 Hörl WH: Removal of high-molecular-weight toxins; in Shaldon S, Koch KM (eds): Polyamide – The evolution of a synthetic membrane for renal therapy. Contrib Nephrol. Basel, Karger, 1992, vol 96, pp 1–138.
11 Jorstad S, Smeby LC, Balstad T, Wideroe TE: Generation and removal of anaphylatoxins during hemofiltration with five different membranes. Blood Purif 1988;6:325–342.
12 Kierdorf H, Melzer H, Riehl J, Heintz B, Weyhenmeyer S, Sieberth HG: Kontinuierliche Hämofiltration beim Multiorganversagen – Elimination von Proteinspaltprodukten und Einfluss auf die proteolytische Aktivität. Intensivmed Notfallmed 1990;27:346–354.
13 Staubach KH, Rau HG, Kooistra A, Schardey HM, Hohlbach G, Schildberg FW: Can hemofiltration increase survival time in acute endotoxemia? A porcine shock model. Prog Clin Biol Res 1989; 308:821–830.
14 Stein B, Pfenninger E, Grünert A, Schmitz JE, Hudde M: Influence of continuous haemofiltration on haemodynamics and central blood volume in experimental endotoxic shock. Intesive Care Med 1990;16:494–502.
15 Storck M, Hartl WH, Zimmerer E, Inthorn D: Comparison of pump-driven and spontaneous haemofiltration in postoperative acute renal failure. Lancet 1991;337:452–461.

Dr. med. Ludwig Blinzler, Klinik für Anästhesiologie und operative Intensivmedizin am Klinikum Nord der Stadt Nürnberg, Flurstrasse 17, D-90419 Nürnberg (Germany)

Sieberth HG, Stummvoll HK, Kierdorf H (eds): Continuous Extracorporeal Treatment in Multiple Organ Dysfunction Syndrome. Contrib Nephrol. Basel, Karger, 1995, vol 116, pp 121–129

Nonrenal Indications for Continuous Hemofiltration Therapy in Patients with Normal Renal Function?

Wilfred Druml

Medizinische Klinik III, Abteilung für Nephrologie, Wien, Österreich

Shortly after the introduction of continuous arteriovenous hemofiltration into clinical medicine it was hypothesized that by continuous treatment modalities aside from maintaining fluid and electrolyte balance, from correction of metabolic acidosis and the elimination of protein degradation products (i.e. aside from 'renal replacement'), other effects can be achieved by the elimination of putative 'mediators' involved in pathophysiology of sepsis and multiple organ failure syndrome (MOFS) [1–3].

Since these original suggestions, numerous uncontrolled and retrospective case series have been published and it was claimed that continuous hemofiltration might present a novel therapeutic approach in several disease states even in absence of renal dysfunction, such as sepsis/septic shock ('systemic inflammatory response syndrome' (SIRS)), multiple organ dysfunction syndrome (MODS), adult respiratory distress syndrome (ARDS), and necrotizing pancreatitis, but also in many other clinical situations (for examples, see table 1) [1–15].

Nevertheless, in spite of ongoing enthusiasm and a constantly increasing number of clinical reports, only few experimental studies have tried to define an 'extra' effect of continuous therapies beyond renal replacement and controlled clinical trials are mostly lacking.

Experimental Studies

Only five experimental studies have tried to assess the impact of continuous hemofiltration on various physiological parameters and survival in animal models of sepsis and endotoxinemia (table 2). Because of shortcomings in study design

Table 1. Nonrenal indications for continuous hemofiltration

Group (first author)	Year	Ref.	Study design	Indication	Parameters of evaluation
Gotloib	1984	1	uncontrolled retrospective	ARDS	hemodynamics gas exchange
Coraim	1985	3	prospective controlled	postop. cardiogen. shock, ± ARDS	hemodynamics gas exchange
Ossenkoppelle	1985	5	retrospective	septic shock	hemodynamics
Rau	1987	6	case series	circulatory shock of various origin	hemodynamics gas exchange
Golper	1988	7	case reports	intoxication	drug removal
Larner	1988	8	case reports	hyponatremia	serum sodium
Davenport	1989	9	retrospective case series	hepatic failure	encephalopathy intracran. pressure
Barzilay	1989	10	retrospective	MODS	hemodynamics
Garcia	1991	11	non-randomized	ARDS	hemodynamics gas exchange
Blinzler	1991	12	uncontrolled retrospective	necrotizing pancreatitis	survival gas exchange
Consentino	1991	13	prospective randomized	ARDS	gas exchange survival
Riegel	1994	14	prospective randomized	trauma	elastase, CRP hemodynamics, O_2 transport
Sander	1994	15	prospective randomized	sepsis	mediators

For explanations, see text.

and the low comparability with the clinical situation, the practical relevance of these studies is limited:

Staubach et al. [16] used a pig model of endotoxic shock to evaluate the effect of hemofiltration (maximally 6 h?) on hemodynamics, gas exchange, survival and serum levels of arachidonic acid metabolites. The specifics of the hemofiltration procedure are not mentioned. As far as can be derived from the figures, no significant differences could be demonstrated between the filtered group and nonfiltering control animals in any of the parameters evaluated. Nevertheless, the authors conclude that hemofiltration might decrease thromboxane and 6-keto-PGF levels and 'counteract some toxic effects of shock mediators'.

Table 2. Hemofiltration in sepsis endotoxic shock – experimental studies

Group (first author)	Year	Ref.	Model of sepsis	Blood flow ml/min	Filtration rate ml/h	Duration of HF min	Non-filtering controls	Parameters of improvement
Staubach	1989	16	pig – endotoxin	?	?	?	yes	PVR, MAP VO_2, Tx-B_2
Gomez	1990	17	dog – *E. coli*	3,000	500	180	no hypotensive controls	SV LV contractility
Stein	1990, 1992	18, 19	pig – endotoxin	50	600	240	yes	PAP, PCWP, PVR, pulm. compliance survival?
Grootendorst	1992	22, 23, 25	pig – endotoxin	300	6,000	240	yes + reinfusion	CO, MAP, LWSW, RVSW, RVEF
Lee	1993	24	pig – *S. aureus*	50–150	8.3–16.7 (?) (500–1,000?)	360	yes + reinfusion	survival time MAP

For explanations, see text.

Gomez et al. [17] investigated the impact of hemofiltration during maximally 180 min with a blood flow of 2,000–3,000 ml/min (!) and a total filtrate volume of 1,500–2,000 ml (!) on hemodynamics and left ventricular contractility in dogs during *Escherichia coli* infusion. Aside from a minimal increase in stroke volume, no in vivo alterations in hemodynamics could be identified. Sepsis-induced decrease in myocardial contractility, as measured by subendocardially positioned crystal transducers, was claimed to be reversed by hemofiltration. No nonfiltering controls were investigated.

Stein et al. [18, 19] studied hemodynamics, central blood volume, extravascular lung water and lung mechanics in swine subjected to endotoxic shock during 4 h of hemofiltration (total filtered volume 2,400 ml). Pulmonary and systemic vascular resistance, pulmonary arterial pressure and left ventricular filling pressures were improved as was pulmonary compliance. In support of clinical observations, extravascular lung water remained unchanged during hemofiltration [20, 21]. A nonfiltering control group was not included so unspecific effects to extracorporeal circulation could not be discriminated.

Grootendorst et al. [22, 23], again using a pig model of endotoxic shock, studied the effect of a zero-balanced hemofiltration during 240 min with a filtration volume of 6,000 ml/h (!) in comparison with a nonfiltering control group on hemodynamic parameters. Mean arterial pressure, cardiac output, left and right ventricular stroke work, and right ventricular ejection fraction were improved, whereas – in sharp contrast to the studies performed by Stein et al. [18, 19] – pulmonary circulation remained unaffected. An independent effect of the excessive amount of lactate infused was not assessed.

Lee et al. [24] evaluated the impact of 'continuous' arteriovenous hemofiltration in a swine model of *Staphylococcus aureus* septicemia on survial time during maximally 6 h of filtration. The authors claim to demonstrate an improvement of survival time in direct correlation with the filtration fraction of the hemofiltration system. The differences in survival time (53.0 ± 10.2 and 70 ± 3.8 h (SEM)) for filtration fractions of 5.5 and 33.4% respectively certainly is neither statistically nor physiologically significant or relevant.

In some of those studies the experiments were extended to the assessement of in vitro effects of the filtrate generated on heart muscle preparation [17] or by reinfusion of the filtrate into control animals [24, 25]. The filtrate generated from septic animals inhibited myocardial contractility and decreased survival when infused into intact animals. Certainly, little information on the physiologic importance or the effectivity of an extracorporeal system can be deduced from such experiments.

Taken together, these studies illustrate the profound confusion that has arisen around continuous replacement therapies. Obviously, none of the available experimental studies simulates 'continuous' therapy as the extracorporeal perfusion time was only 3 h to a maximum of 6 h and, at best, may reflect 'intermittent' hemofiltration. Neither blood flow rates of 3,000 ml/min nor filtration rates of 6,000 ml/h have any relevance to the clinical situation. Results were mostly inconsistent, physiologically of little relevance and often contradictory. With any possible production kinetics of a given endogenous 'factor' it is naive to expect that the mere filtration of 1.5–3 liters during the early 360 min of a spetic process will exert any substantial impact on prognosis. It is very unfortunate that an experimental study following a study design more relevant to the clinical situation is not available.

Evidence for the Elimination of 'Mediators'

Since the original description on the elimination of 'myocardial depressant factor' during hemofiltration by Coraim et al. [3] (which could not be reproduced by other groups), multiple investigations have dealt with the elimination of 'mediators' such as tumor necrosis factor (TNF), Interleukin (IL)-1, IL-6, IL-8, platelet-activating factor (PAF) or complement factors during continuous hemofiltration [26–29].

An example for the discussion of the clinical relevance of the elimination of a putative factor is TNF. In some investigations it was suggested that by hemofiltration TNF is eliminated because of the small molecular weight of 17 kD [26]. Later it became clear that the majority of TNF is present in the circulation as a trimer so it is not filtered but – because in fact TNF can be detected in the filtrate – potentially is generated in the extracorporeal circuit by cellular activation. Nevertheless, the same author evaluated the 'clearance' of TNF and IL-1 by hemofiltration in septic patients [27]. In spite of the fact that the total elimination corresponded to a clearance of 36 liters/day only (the generation in the extracorporeal circuit was not evaluated) and there was no effect on plasma concentrations of these factors, the authors come to the more than speculative conclusion that the elimination of mediators may 'attenuate the progression of MODS'.

Certainly, many circulating peptides such as insulin, catecholamines but also 'mediators' such as ILs are filtered during hemofiltration [29]. To assess any potential physiological importance, one has to consider production kinetics, the synthesis rate, the whole body tunover, plasma half-life and volume of distribution. Peptide hormones, such as insulin or catecholamines, have a high fractional extraction rate, but nobody would claim that continuous hemofiltration induces diabetes mellitus or aggravates hemodynamics instability [28]. One should resist to ascribe the well-defined advantages of continuous renal replacement modalities to ill-defined elimination of unknown mediators.

When evaluating any potential elimination of factors in a patient with preserved renal function, one should keep in mind that kidneys are an extremely efficient organ of degradation of multiple proteins including inflammatory mediators. The efficiency of this endogenous system is much higher than any possible extracorporeal elimination by hemofiltration!

Clinical Experience

Numerous clinical studies on potential nonrenal indications of continuous treatment modalities have been published during the past decade (table 1) [1–15]. In most instances these comprised retrospective case series without adequate definition of parameters of assessment or end points [e.g. 30]. Many authors, nevertheless, come to firm conclusions in respect to the effectivity of this therapeutic approach [for review, see 31].

In the only major controlled trial on continuous hemofiltration in acute renal failure (ARF), Storck et al. [32] have compared spontaneous arteriovenous with pump-assisted continuous hemofiltration in patients with postoperative ARF. The authors claim that the survival rate is correlated with the filtered volume per day and have suggested – and this is important for the discussion of nonrenal effects of hemofiltration – that this is mediated by the elimination of undefined factors. However, an inadequate renal replacement therapy with a filtration volume of <7.5 liters/day (!) was compared with a possibly sufficient treatment (filtration volume >15 liters/day). It is not surprising that mortality in the group with the lower filtration volume was higher than in any other study on the prognosis of ARF.

Results from the few prospective, randomized and controlled studies on a nonrenal application of continuous hemofiltration are mostly available as abstracts or preliminary communications in meeting proceedings [13–15].

Consentino et al. [13] compared the therapy of ARDS with (n = 9) and without (n = 6) additional CAVH. Gas exchange was even better in the control group, but there was a trend (p = 0.29) of enhanced survival in the treatment group (5/9 vs. 1/6).

Riegel and co-workers [14] conducted a controlled study comparing continuous venovenous hemofiltration versus no filtration in multiple trauma patients with normal renal function. There was no impact on survival but mortality was negligible in both groups. As a possible consequence of prolonged blood-membrane interactions, C-reactive protein significantly rose in the hemofiltration group. Cardiac output and oxygen delivery were decreased, systemic vascular resistance increased during hemofiltration. This 'reversal of hyperdynamic circulation' contradicts the aim of optimizing oxygen delivery in intensive care patients.

Lastly, Sander et al. [this volume] compared continuous hemofiltration versus standard control treatment in patients with early sepsis [15]. Except for marginal alterations in plasma TNF and IL-6 concentrations, no clinically relevant alterations have been observed during hemofiltration and no differences in prognosis were identified.

Other Effects of Extracorporeal Treatment Modalities

Positive effects: Aside from the potential filtration (and/or diffusion), molecules can also be adsorbed on the artificial surface, can be trapped in the gel layer, can be activated or inhibited. Adsorption may become relevant for endotoxins, a process which, however, is saturable [33]. Additional advantages of continuous hemofiltration include restoration of body fluid spaces and especially, extracellular volume, hemodynamic stability, reversal of a hyperpyretic state. The infusion of large amounts of organic buffer, such as lactate or acetate, may have profound effects on energy metabolism and myocardial function.

Negative effects: One should not forget that continuous extracorporeal treatment modalities are associated with multiple, potentially serious disadvantages (table 3). The prolonged blood-membrane interaction activates obligatory phe-

Table 3. Potential disadvantages, side effects and complications of continuous hemofiltration

Prolonged blood-membrane interaction with obligatory phenomena of bioincompatibility
Induction of a chronic inflammatory reaction
Activation of protein catabolism
Substrate losses (glucose, amino acids, etc.)
Loss of hormones
Heat loss
Unphysiologic load of organic anions (lactate, acetate)
Need for prolonged anticoagulation (risk of bleeding)
Decrease in oxygen delivery
Catheter-associated complications
Risk of infections
Immobilization of the patient
Time-consuming procedure
High costs

nomena of bioincompatibility and may induce a chronic inflammatory response with multiple adverse effects which, possibly, can enhance protein catabolism and aggravate SIRS [34, 35]. The need for continuous anticoagulation, the necessity of large-bore catheters, the obligatory loss of substrates such as amino acids and small peptides, unphysiologic load with substrates such as lactate, acetate, and heat loss all may negatively affect the clinical course of a specific patient. The associated decrease in oxygen delivery may favor the development of a regional oxygen deficit. Last but not least, continuous hemofiltration is time-consuming and associated with high costs.

Conclusions

In spite of an obvious lack of sound experimental and/or clinical data, continuous hemofiltration has been recommended – as 'nonrenal' indications – in the treatment of various clinical conditions with preserved renal function and especially, those associated with SIRS and/or MODS. However, at present there is no evidence whatsoever that any patient suffering from one of the disease states mentioned in table 1 and not grossly impaired renal function will profit from the institution of continuous hemofiltration and that this will prevent the development of SIRS or organ dysfunctions when used prophylactically.

One should keep in mind that continuous hemofiltration is associated with multiple side effects and potential complications (table 3). From an ethical point of view this makes the institution of this therapeutic approach not justified in patients with maintained renal function, *except* in carefully designed clinical studies.

A modification of 'mediator system' by hemofiltration, the extracorporeal elimination (filtration, adsorption, etc.) of molecules involved in the pathophysiology of various disease states, such as 'granulocyte-inhibiting proteins' [36] and especially endotoxin is a new and stimulating approach in the therapy of SIRS and MODS but what is urgently needed are carefully designed experimental and clinical studies, a better understanding and definition of substances involved and a careful assessment of the clinical relevance. Future developments will be boosted by the introduction of newer membrane materials with enhanced adsorptive properties and with higher and better defined cut-offs for specific molecules.

References

1 Gotloib L, Barzilay E, Shustak A, Lew A: Sequential hemofiltration in nonoliguric high capillary permeability pulmonary edema of severe sepsis: Preliminary report. Crit Care Med 1984;12:997–1000.

2 Goitlob L, Barzilay E, Shustak A, Wais Z, Jaichenko J, Lev A: Hemofiltration in septic ARDS. The artificial kidney as an artificial endocrine lung. Resuscitation 1986;13:123–132.

3 Coraim F, Pauser G, Stellwag F, Werner T, Ziegler W: Positive modification of hemodynamics in postcardiac surgery patients by hemofiltration. Improved method for the demonstration of myocardial depressant factor in hemofiltrate. Anaesthesist 1985;34:236–240.

4 Coraim FJ, Coraim HP, Ebermann R, Stellwag FM: Acute respiratory failure after cardiac surgery: Clinical experience with continuous arteriovenous hemofiltration. Crit Care Med 1986;14:714–718.

5 Ossenkoppelle GJ, van der Meulen J, Bronsveld W, Thijs LG: Continuous arteriovenous hemofiltration as an adjunctive therapy for septic shock. Crit Care Med 1985;13:102–104.

6 Rau HC, Staubach KH, Hohlback C, Klingler W: The continuous arteriovenous hemofiltration in shock. Prog Clin Biol Res 1987;236B:241–247.

7 Golper TA, Bennet WM: Drug removal by continuous arteriovenous hemofiltration: A review of the evidence in poisoned patients. Med Toxicol 1988;3:341–349.

8 Larner AJ, Vickers CR, Adu D, Buckels JAC, Elias E, Neuberger J: Correction of severe hyponatremia by continuous arteriovenous hemofiltration. Br Med J 1988;297:1514–1515.

9 Davenport A, Will EJ, Davison AM, Swindells S, Cohen AT, Miloszewski KJA, Losowsky MS: Changes in intracranial pressure during hemofiltration in oliguric patients with garde IV hepatic encephalopathy. Nephron 1989;53:142–146.

10 Barzilay E, Kessler D, Berlot G, Gullo A, Geber D, Zeev IB: Use of extracorporeal supportive techniques as additional treatment for septic-induced multiple organ failure patients. Crit Care Med 1989;17:634–638.

11 Garcia F, Todor R, Scalea T: Continuous arteriovenous hemofiltration-countercurrent dialysis in acute respiratory failure. J Trauma 1991;31:1277–1285.

12 Blinzler L, Hausser J, Bödecker H, Zaune U, Martin E, Gebhardt Ch: Conservative treatment of severe necrotizing pancreatitis using early continuous venovenous hemofiltration. Contrib Nephrol, Basel, Karger, 1991, vol 93, pp 234–236.

13 Cosentino F, Paganini E, Lockrem J, Stoller J: Continuous arteriovenous hemofiltration in the adult respiratory distress syndrome. A randomized trial; in Sieberth HG, Mann H, Stummvoll HK (eds): Continuous Hemofiltration. Contrib Nephrol. Basel, Karger, 1991, vol 93, pp 94–97.

14 Mazi I, Ziegenfuss Th, Rose St, Bauer M, Riegel W: Einfluss der kontinuierlichen venovenösen Hämofiltration auf die posttraumatische Entzündungsreaktion. Beitr Anästh Notfallmed 1994;7: 52–67, 87–99.

15 Sander A, Armbruster W, Sander B, Philipp Th, Schäfer Ch, Thürauf N, Lange R: Elimination of TNFα and IL-6 continuous hemofiltration in the early phase of sepsis. 3rd Int Congr Continuous Hemofiltration, Vienna 1994.
16 Staubach KH, Hau HG, Kooistra A, Schardey HM, Hohlbach G, Schildberg FW: Can hemofiltration increase survival time in acute endotoxinemia – A porcine shock model. Prog Clin Biol Res 1989;308:821–826.
17 Gomez A, Wang R, Unruh H, Light RB, Bose D, Chau T, Correa E, Mink S: Hemofiltration reverses left ventricular dysfunction during sepsis in dogs. Anesthesiology 1990;73:671–685.
18 Stein B, Pfenninger E, Grünert A, Schmitz JE, Hudde M: Influence of continuous haemofiltration on haemodynamics and central blood volume in experimental endotoxic shock. Intensive Care Med 1990;16:494–499.
19 Stein B, Pfenninger E, Grünert A, Schmitz JE, Deller A, Kocher F: The consequences of continuous hemofiltration on lung mechanisms and extravascular lung water in a porcine endotoxic shock model. Intensive Care Med 1992;17:293–298.
20 Laggner AN, Lenz K, Grimm G, Sommer G, Gössinger H: Hämofiltration zur Reduktion des Lungenwassers bei ARDS. Schweiz Med Wochenschr 1987;117:445–449.
21 Laggner AN, Druml W, Lenz K, Schneeweiss B, Grimm G: Influence of ultrafiltration/hemofiltration on extravascular lung water. Contrib Nephrol. Basel, Karger, 1991, vol 93, pp 65–70.
22 Grootendorst AF, van Bommel EFH, van der Hoven B, van Leengoed LAMG, van Osta GALM: High-volume hemofiltration improves hemodynamics of endotoxin-induced shock in the pig. J Crit Care 1992;7:67–75.
23 Grootendorst AF, van Bommel EFH, van der Hoven B, van Leengoed LAMG, van Osta GALM: High-volume hemofiltration improves right ventricular function in endotoxin-induced shock in the pig. Intensive Care Med 1992;18:235–240.
24 Lee PA, Matson JR, Pryor RW, Hinshaw LB: Continuous arteriovenous hemofiltration therapy for *Staphylococcus aureus*-induced septicemia in immature swine. Crit Care Med 1993;21:914–924.
25 Bommel EFH, van Grootendorst AF, van Lenngoed LAMG: The influence of high-volume hemofiltration on hemodynamics in porcine endotoxic shock. Blood Purif 1992;10:88–89.
26 Bellomo R, Tipping P, Boyce N: Tumor necrosis factor clearance during veno-venous hemodiafiltration in the critically ill. Trans Am Soc Intern Organs 1991;37:322–323.
27 Bellomo R, Tipping P, Boyce N: Continuous veno-venous hemofiltration with dialysis removes cytokines from the circulation of septic patients. Crit Care Med 1993;21:522–526; see also letters 1994;22:715–721.
28 Bellomo R, McGrath B, Boyce N: In vivo catecholamine extraction during continuous hemodiafiltration in inotrope-dependent patients. Trans Am Soc Intern Organs 1991;37:324–325.
29 Kierdorf HR: Einfluss der kontinuierlichen Hämofiltration auf Proteinkatabolismus, Mediatorsubstanzen und Prognose des akuten Nierenversagens in Multiorganversagen; thesis, Aachen 1993.
30 Bagshaw ON T, Anaes FRC, Hutchinson A: Continuous arteriovenous haemofiltration and respiratory function in multiple organ systems failure. Intensive Care Med 1992;18:334–338.
31 Grootendorst AF, van Bommel EFH: The role of hemofiltration in the critically ill intensive care unit patient. Present and future. Blood Purif 1993;11:209–223.
32 Storck M, Hartl WH, Zimmerer E, Inthorn D: Comparison of pump-driven and haemofiltration in postoperative acute renal failure. Lancet 1991;337:452–455.
33 Dinarello CA, Lonnemann G, Maxwell R, Shaldon S: Ultrafitration to reject human interleukin-1-inducing substances derived from bacterial cultures. J Clin Microbiol 1987;25:1233–1238.
34 Hakim RM: Clinical implications of hemodialysis membrane biocompatibility. Kidney Int 1993; 44:484–494.
35 Gutierrez A, Alverstrand A, Bergström J: Membrane selection and muscle protein catabolism. Kidney Int 1992;42(suppl 38):S86–S90.
36 Haag-Weber M, Mai B, Cohen G, Hörl WH: GIP and DIP: A new view of uremic toxicity. Nephrol Dial Transplant 1994;9:346–347.

Prof. Dr. Wilfred Druml, Medizinische Klinik III, Abteilung für Nephrologie,
Währinger Gürtel 18–20, A–1090 Wien (Austria)

Sieberth HG, Stummvoll HK, Kierdorf H (eds): Continuous Extracorporeal Treatment in Multiple Organ Dysfunction Syndrome. Contrib Nephrol. Basel, Karger, 1995, vol 116, pp 130–135

Low-Dose Heparinization for Anticoagulation in Intensive Care Patients on Continuous Hemofiltration

N. Gretz[a], *M. Quintel*[b], *M. Ragaller*[b], *W. Odenwälder*[a], *H.J. Bender*[b], *P. Rohmeiss*[a], *M. Strauch*[a]

Divisons of
[a] Nephrology and
[b] Anesthesiology, Klinikum Mannheim, University of Heidelberg, Mannheim, Germany

Continuous pump-driven hemofiltration is more and more frequently used in intensive care patients [1]. As this treatment modality requires anticoagulation in the therapeutic range, a considerable bleeding risk exists. This is especially true in patients recovering from polytrauma or major surgery. Furthermore, in patients exhibiting coagulation disorders, therapeutic heparinization may cause unexpected hemorrhagic events. An alternative to therapeutic heparinization is either to administer no heparin at all or to reduce heparin dosage to a degree that no prolongation of prothrombin time (PTT) and activated clotting time (ACT) occurs. This, however, may result in clotting of the filter. Consequently, other approaches in order to overcome systemic heparinization have been applied, like regional heparinization, citrate anticoagulation or prostaglandin and gabexate administration. Any of these approaches, however, has serious disadvantages resulting in an additional risk for the patient. Another though time-consuming approach is to flush the system at short time intervals.

The aim of this paper is to describe an approach allowing continuous hemofiltration/hemodialysis with low-dose heparinization.

Methods

The underlying idea of our approach is to fix heparin on the filter surface by modifying the prerinsing procedure and to reduce blood-air contact. In addition, we change the hemofilters every 24 h to avoid filter clotting and clearance/adsorption loss.

For hemofiltration we use either a polysulfone hemofilter (AV600®; Fresenius, FRG) or a hemophane filter (BLX 16 GW®; Nikiso, Japan). In a preliminary study we could demonstrate that heparin was absorbed by both filters [2]. The filters are rinsed in a recirculation mode with 2 liters of normal saline containing 20,000 units of heparin. The recirculation is performed for 1 h. Then the circuit is flushed with 1 or 2 liters of normal saline to wash out the unbound heparin. Thereafter the patient is connected to the system. In the concentration of heparin used for prerinsing, the amount of heparin absorbed ranges from 20 to 90%. Furthermore, no significant heparin desorption occurred during short-time hemodialysis sessions.

As blood-air contact results in a marked activation of the clotting system, we constructed a new venous blood chamber (Hospal BSM-V 148®) by extending the inlet blood line (fig. 1). After filling the circuit with blood, the blood level in the venous blood chamber is allowed to rise about 1 cm above the inlet level of the venous blood line. During the course of the hemofiltration session, blood sedimentation occurs leaving a plasma layer on top of the circulating blood. Thus this layer acts as an air lock. Extending the venous inlet line also prevents the blood from falling through the air into the blood chamber. Thus an additional blood-air contact is inhibited.

In the arterial blood chamber we prevent the blood-air contact by infusing a pH-neutral predilution solution at a rate between 1 and 2 liters. Thus there is a layer of infusion solution functioning as an air lock (fig. 1).

As predilution solution we use a pH-neutral, lactate (3 mmol/l) and bicarbonate-buffered (30 mmol/l) solution (HF43®; Fresenius). We prefer this hemofiltration solution over other hemofiltration solutions as it is pH-neutral and does not contain unphysiologically high lactate concentrations. The continuous hemofiltration or hemodialysis is performed with a BSM 22® or BSM 32® (Hospal, FRG) at a blood flow rate of 100 ml/min. Filters are changed every 24 h in order to prevent clotting and thus to minimize the treatment-induced blood loss. Furthermore, we speculate that the loss of clearance and/or adsorption for a number of substances, like inflammatory mediators, may be reduced.

Heparinization is controlled by ACT and PTT. ACT is determined by using the Hemotec® analyzer, while PTT is assessed by standard laboratory methods. In addition, fibrinogen determination and platelet count are performed, as the 'missing' heparin effect could result in fibrinogen degradation or platelet loss.

Results

Twenty-seven intensive care patients were treated with 218 prerinsed filters and the modified blood tubing systems. The median age of the patients was 49 ± 18 years and body weight averaged 85 ± 21 kg. On average, 11,000 units of heparin/day were administered. The mean amount of heparin was 135 units/kg/day. No bleeding or thromboembolization occurred. In 12% of the filters, clotting was observed. The number and size of the clots, however, in the blood chambers (either venous arterial) was less pronounced with the new approach than what we had previously observed.

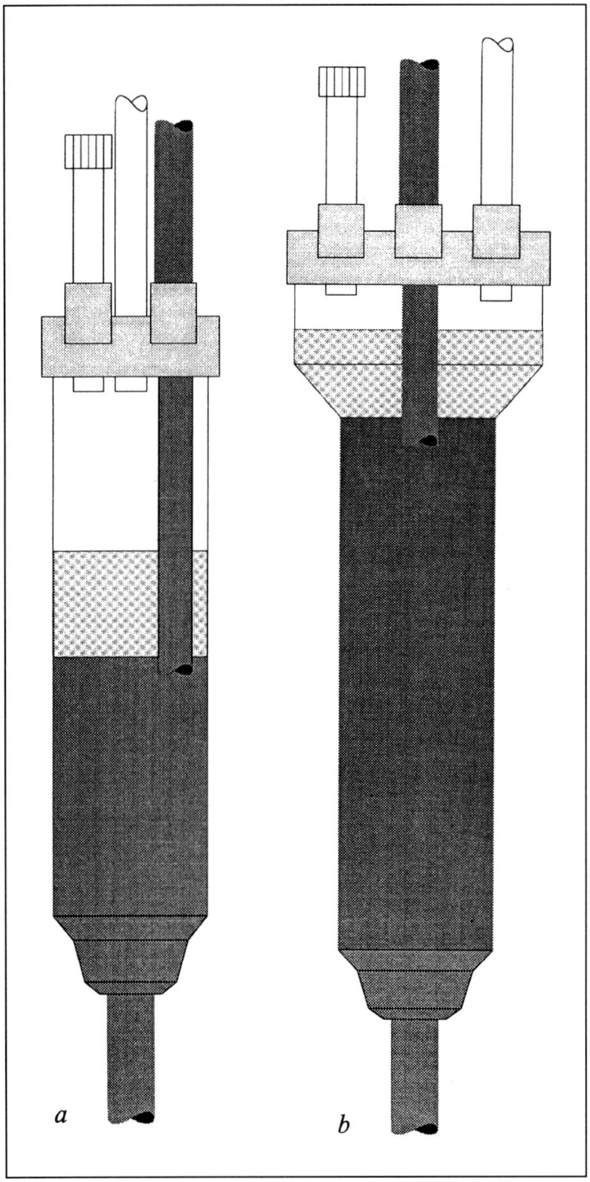

Fig. 1. Modified filling of the arterial blood chamber (*a*) and the modified blood inlet of the venous blood chamber (*b*).

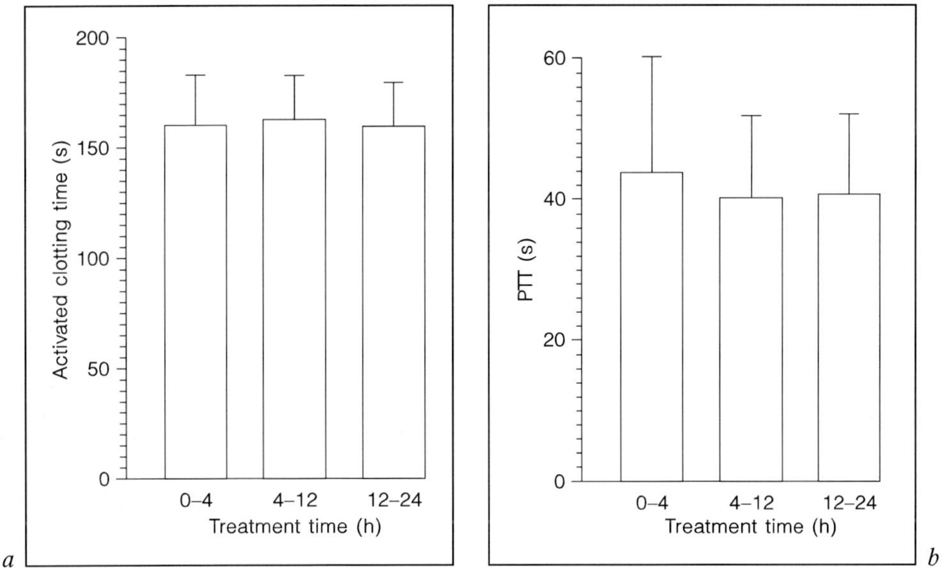

Fig. 2. Course of ACT (*a*) and PTT (*b*) during low-dose heparinization.

Due to the prerinsing with high heparin concentration, one could have speculated that a significant desorption of heparin might occur resulting in a rise of ACT or PTT. Such a rise was not noted (fig. 2). On the contrary, there was a slight decrease over time indicating a slight coagulation activation. If a significant coagulation activation would have occurred, a decrease of platelet count and fibrinogen concentration would have resulted. However, only a slight decrease in platelet count was noted (0–4 h: 119,000 ± 110,000; 4–12 h: 117,000 ± 90,000; 12–24 h: 100,000 ± 66.000). In contrast, the fibrinogen levels rose slightly from 0–4 h: 476 ± 112 mg/dl to 4–12 h: 487 ± 100 mg/dl and reaching finally 12–24 h: 511 ± 112 mg/dl.

Discussion

In this paper we describe an approach to overcome the drawbacks of the usual high-dose anticoagulation by: (1) adsorbing heparin to the filter membrane by means of a modified prerinsing procedure and administering low-dose heparin, and (2) by reducing the blood-air contact by means of air locks in the arterial and venous blood chamber. This approach enabled us to perform continuous hemofiltration sessions with low-dose heparinization.

Recently we could demonstrate that by using the proposed prerinsing procedure a considerable amount of heparin can be fixed in vitro to the membrane of hemodializers [2]. It is of note that our prerinsing procedure only works at high heparin concentrations (10 IU/ml) and the efficacy seems to be time-dependent. Comparable prerinsing procedures had been described. Usually, however, much lower heparin concentrations were administered. Furthermore, for some membrane types a high rate of desorption had been described [3]. We speculate that due to the high concentration the heparin molecule sticks upright on the filter surface. This results in high anticoagulation activity as it is known from preheparinized ECMO systems. The exact mechanism by which the heparin adsorption occurs is unknown.

In the present paper we confirm that the adsorbed amount of heparin is sufficient to perform continuous hemofiltration without high-dose/therapeutic anticoagulation. Furthermore, we can demonstrate that no relevant heparin desorption occurs during the hemofiltration sessions, as ACT and PTT remained unchanged.

We feel that reducing the blood-air contact is of paramount importance. The relevance of blood-air contact is indicated by the frequent finding of clots in the venous blood chamber. A reduction in blood-air contact was achieved by modifying the blood inlet into the venous blood chamber and by elevating blood level there. This resulted in a plasma layer functioning as an 'air lock'. The efficiency of this approach was revealed by a reduction of the number and size of clots in the venous chamber. The filling of the arterial blood chamber with predilution solution also reduced the blood-air contact.

The impact of large volume predilution resulting in a fluid layer in the arterial blood chamber is debatable as doing so results also in a fairly high filtration volume, which could stimulate coagulation in the filter. Furthermore, using predilution has the disadvantage that larger fluid volumes are needed to obtain the same treatment efficiency [4, 5].

Changing filters every day seems to be an expensive approach. By the frequent changes, clotting and consecutive blood loss is minimized. Thus the risk of infection is reduced. Furthermore, the potential loss of clearance over time can be mimimized. An additional reason for changing filters frequently is that if 'mediators' are adsorbed to the filter surface, there could be a saturation of the membrane surface. Thus after a certain period of time, no further adsorption may occur. On the contrary, even a desorption may result. This is another reason why we prefer large filter surfaces over the normally used hemofilters. On the other hand, using large filter surfaces reduces the blood flow/fiber favoring premature filter clotting.

In summary, we describe an approach of continuous hemofiltration/hemodialysis suitable for use in patients having high risk of bleeding. Our modified approach (prerinsing with heparin and reducing blood-air contact) allows to perform continuous hemofiltration/hemodialysis with low-dose heparinization leading to no risk of bleeding.

References

1 Grootendorst AF, van Bommel EFH: The role of hemofiltration in the critically-ill intensive care unit patient: Present and future. Blood Purif 1993;11:209–223.

2 Gretz N, Schulz I, Bowry S, Vienken J, Hafner G, Odenwälder W, Quintel M, Rohmeiss P: Heparin-free dialysis following a new rinsing procedure (submitted).

3 Schmitt E, Holtz M, Klinkmann H, Esther G, Courtney J: Heparin binding and release properties of DEAE-cellulose membranes. Biomaterials 1983;4:309–313.

4 David S, Cambi V: Hemofiltration: Predilution versus postdilution; in Shaldon S, Koch K (eds): Polyamide – The evolution of a synthetic membrane for renal therapy. Contrib Nephrol. Basel, Karger, 1992, vol 96, pp 77–85.

5 Kaplan AA: Predilution versus postdilution for continuous arteriovenous hemofiltration. Trans Am Soc Artif Organs 1985;31:28–32.

N. Gretz, MD, Divison of Nephrology, University of Heidelberg, Klinikum Mannheim, D-68167 Mannheim (Germany)

Sieberth HG, Stummvoll HK, Kierdorf H (eds): Continuous Extracorporeal Treatment in Multiple Organ Dysfunction Syndrome. Contrib Nephrol. Basel, Karger, 1995, vol 116, pp 136–139

Use of Prostacyclin in Patients with Continuous Hemofiltration after Open Heart Surgery

Karl-Friedrich Klotz, Horst Gellersen, Thomas Brzelinski, Beate Sedemund-Adib, Hartmut Gehring

Department of Anesthesiology, Medical University of Lübeck, Germany

Usually, a patient who underwent open heart surgery will be discharged from the intensive care unit (ICU) the first postoperative day. However, the risk of serious postoperative complications is high in this surgical speciality. The incidence of acute renal failure is described to get up to 30% [1] and the patients concerned suffer an overall mortality of 67%. Treatment of postoperative anuria can be done by replacing renal function by continuous hemofiltration. The application of this method requires high anticoagulation to guarantee blood flow without clotting through the filter capillaries. Anticoagulation is mostly performed by infusion of heparin, but there are some problems connected with this substance. Heparin-associated thrombocytopenia (HAT) occurs in up to one third of heparin-treated patients [2]. HAT is of special importance in cardiosurgical patients as the plasma and cellular components of the coagulation system may already be disturbed by the extracorporeal circulation, by preoperatively taken drugs (aspirin), or by intraoperative hemodilution. In addition to the antiaggregatory effect of heparin, it seems to be advantageous to inhibit thrombocyte aggregation by prostacyclin [3]. By the clinical observations presented here we sought to gain information about the risks and advantages of the anticoagulatory methods, heparin and prostacyclin, used in our ICU for hemofiltrated patients.

Patients and Methods

From March 1993 to March 1994, we observed 13 patients (10 male, 3 female; age 44–79 years) in our ICU. Those underwent open heart surgery and suffered postoperatively from multiorgan failure including acute renal failure. All patients were anticoagulated by

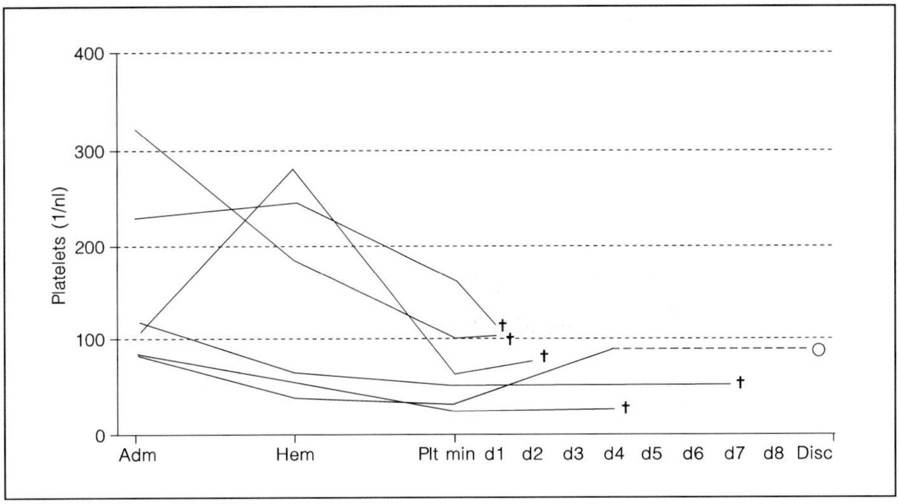

Fig. 1. Platelet count of 6 individual cardiosurgical patients before and during CVVH treatment with heparin. Shown are the courses of platelet counts of each single patient treated with heparin. Adm = Time of postoperative admission to the ICU; Hem = time of start of continuous hemofiltration; Plt min = minimal number of platelets during CVVH; d1, 2, ... = number of days after reaching minimal platelet count, † = patient who died during treatment; ○ = surviving patient.

continuous infusion of heparin to obtain a partial thrombin time (PTT) of 35 s. One to 15 days after admission to the ICU, continuous venovenous hemofiltration (CVVH) was necessary due to acute renal failure. CVVH was performed via a 10-French double-lumen catheter inserted into the femoral vein using Hospal-Multiflow-100 hemofilters and a roller pump system. Heparin infusion rate was increased to achieve a PTT of 45 s. The mean dosage of heparin was 600 U/h. After observation of 6 heparin-treated patients we changed the routine therapeutic regimen. Further, in 7 cases of rapid decrease of platelet count (to about 40/nl), the anticoagulation with heparin was reduced to an average of 350 U/h and supplemented with prostacyclin (Flolan®; Wellcome Co.) which was infused into the arterial line of the CVVH system in a mean dosage of about 20 μg/h.

Results

The course of platelet count in 6 cardiosurgical heparin-treated patients with acute renal failure and continuous hemofiltration is shown in figure 1. During the first postoperative days after admission there is no uniform change of platelet count, but after start of hemofiltration, platelet count of all patients dropped. There is almost no recovery seen in the heparin-treated patients. Five of 6

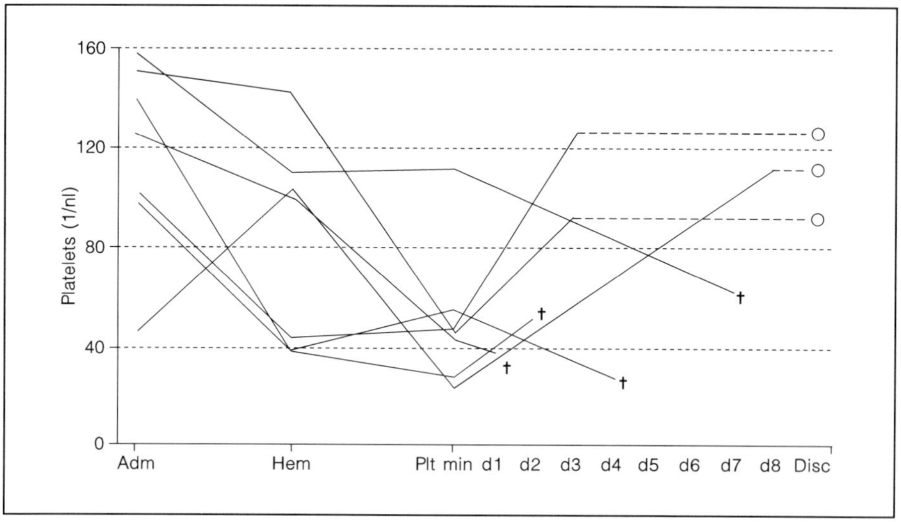

Fig. 2. Platelet count of 7 individual cardiosurgical patients before and during CVVH treatment with prostacyclin. Shown are the courses of platelet counts of each single patient treated with prostacyclin. PGI_2 = Time of start of prostacyclin treatment. For further notes, see legend of figure 3.

observed patients died within the following 7 days of hemofiltration treatment. Only 1 patient (whose platelet count increased) was discharged from the ICU. From these figures the mortality rate is 83%. Following this experience, 7 further patients with multiorgan failure were observed. The course of the platelet count is demonstrated in figure 2. Again, the number of thrombocytes decreased after start of CVVH (to approx. 40/nl). After start of additional prostacyclin treatment, however, the platelet number increased up to the threefold of minimal level in 4 of 7 patients. Three of the patients survived and were discharged from the ICU. From these figures there is a mortality rate of 57%. In figure 3 the mean courses of platelet number of the different groups are shown. Survivors are depicted separately from nonsurvivors. An overall decrease of platelet number is seen after start of CVVH in all groups. But only patients of groups with a recovered number of thrombocytes survived. The overall mortality in our 13 patients was 69% (heparin group 83%, prostacyclin group 57%). Duration of treatment phases is comparable in the two groups.

To summarize: (1) those patients who have developed multiorgan failure suffered from a decrease in thrombocyte numbers; (2) the number of thrombocytes also decreased markedly after starting CVVH; (3) an increasing number of throm-

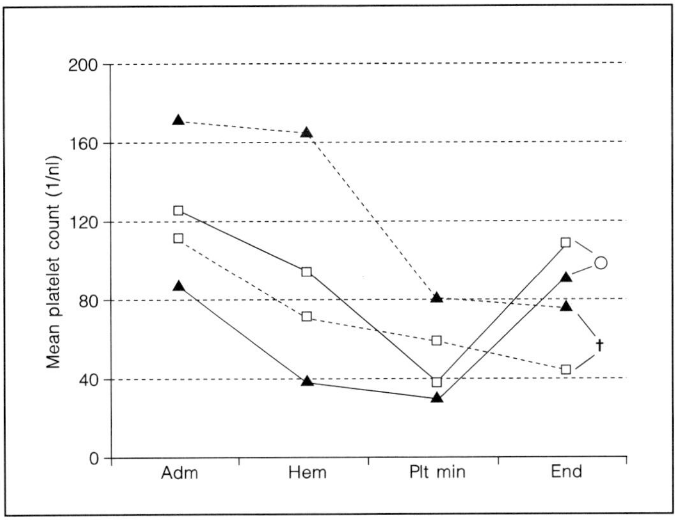

Fig. 3. Mean course of platelet count in groups separated by different anticoagulatory treatment or outcome. —□— = Prostacyclin survivors (n = 3); ---□--- = prostacyclin nonsurvivors (n = 4); —▲— = heparin survivors (n = 1); ---▲--- = heparin nonsurvivors (n = 5). End = Time of end of treatment (discharge or death). For further notes, see legend of figures 1 and 2.

bocytes was indicative for groups of survivors; (4) in some patients treated with prostacyclin the platelet count recovered, and (5) the results of our observation suggest consideration of the early use of prostacyclin.

References

1 Conahan TJ: Complications of cardiac surgery; in Kaplan JA (ed): Cardiac Anesthesia. Philadelphia, Saunders, 1993, pp 1239–1252.
2 Cines DB, Kaywin P, Bina M, Tomaski A, Schreiber AD: Heparin-associated thrombocytopenia. N Engl J Med 1980;303:788–795.
3 Canaud B, Mion C, Arujo A, N'Guyen QV, Paleyrac G, Hemmendinger S, Cazenave JP: Prostacyclin (epoprostenol) as the sole antithrombotic agent in postdilutional hemofiltration. Nephron 1988;48:206–212.

Dr. med. Karl-Friedrich Klotz, Department of Anesthesiology, Medical University of Lübeck, Ratzeburger Allee 160, D–23538 Lübeck (Germany)

Sieberth HG, Stummvoll HK, Kierdorf H (eds): Continuous Extracorporeal Treatment in Multiple Organ Dysfunction Syndrome. Contrib Nephrol. Basel, Karger, 1995, vol 116, pp 140–144

Influencing Hemostasis during Continuous Venovenous Hemofiltration after Acute Renal Failure: Comparison with Intermittent Hemodialysis

Ioannis Stefanidis, Johannes Hägel, Horst Kierdorf, Norbert Maurin

Medical Clinic II (Director: Prof. H.G. Sieberth), University Hospital, Aachen, Germany

In acute renal failure, hemostasis is influenced by uremia, by disseminated intravascular coagulation (DIC) occurring in the context of a multiorgan failure and by renal replacement therapy (heparin, contact with foreign surfaces) [1, 2].

The established therapy method of acute renal failure is intermittent hemodialysis (HD). However, during the last decade a number of continuous treatment methods have been developed [3]. Clinical or experimental studies comparing these new forms of treatment with intermittent HD are missing. In this study we compared the effects of intermittent HD and continuous venovenous hemofiltration (CVVH) on hemostasis after acute renal failure.

Subjects and Methods

We examined 35 patients (25 men and 10 women; mean age 61 years) with acute renal failure of differing genesis, randomized for renal replacement therapy (CVVH 18, HD 17). Patients with venous thrombosis, pulmonary embolism, arterial thrombosis or embolism and patients with artificial mechanical valves at the beginning of treatment were excluded from the study. A polysulfone hemofilter (Ultraflux AV 600®; Fresenius, Oberursel, Germany) was used for CVVH and a high-flux polymethylmethacrylate dialysator (B 1-1.3®; Toray, Tokyo, Japan) was used for HD. The heparin dose was matched to the activated clotting time (ACT > 110 s).

Coagulation tests were performed before renal replacement therapy, at 3-day intervals during treatment and after completion of the therapy. Fibrinopeptide A and thrombin-

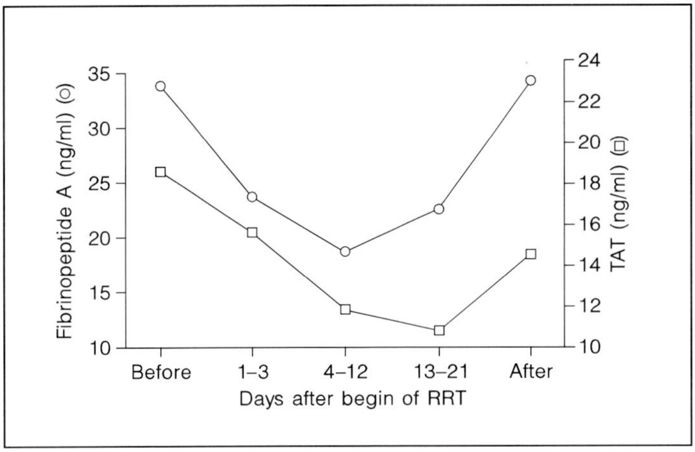

Fig. 1. Coagulation activation markers (fibrinopeptide A; thrombin-antithrombin III complex, TAT) during renal replacement therapy (RRT) both in CVVH and HD patients (n = 35).

antithrombin III complex were used as markers for plasma coagulation and β-thromboglobulin, β-thromboglobulin/creatinine ratio and platelet retention as markers for platelet activation. ELISA was used to determine the plasma concentrations of fibrinopeptide A, β-thromboglobulin (Boehringer, Mannheim, Germany) and thrombin-antithrombin III complex (Behringwerke, Marburg, Germany) [4, 5]. Whole blood was taken and processed immediately with Adeplat-S glass bead filters (Semmelweis, Milan, Italy) to determine Hellem II platelet retention [6]. Platelets were counted before and after filtering; the difference (retention) was given as a percentage (ref. 60–99%).

Results

Before renal replacement therapy, fibrinopeptide A and thrombin-antithrombin III complex were significantly increased, with values of 34 ± 22 ng/ml (ref. < 3.0) and 19 ± 15 ng/ml (ref. 1.0–4.0) respectively. During the course of treatment, fibrinopeptide A and thrombin-antithrombin III did not rise in both treatment groups (fig. 1).

Platelet retention (Hellem II) was reduced prior to renal replacement therapy at 47 ± 22%. During treatment, there was a statistically significant fall in platelet retention to 34 ± 23% ($p < 0.05$) and a distinct rise in the β-thromboglobulin/creatinine ratio from 0.50 ± 0.30 to 0.77 ± 0.50 in both groups (fig. 2). During treatment, platelet retention in the CVVH group was significantly ($p < 0.05$) lower

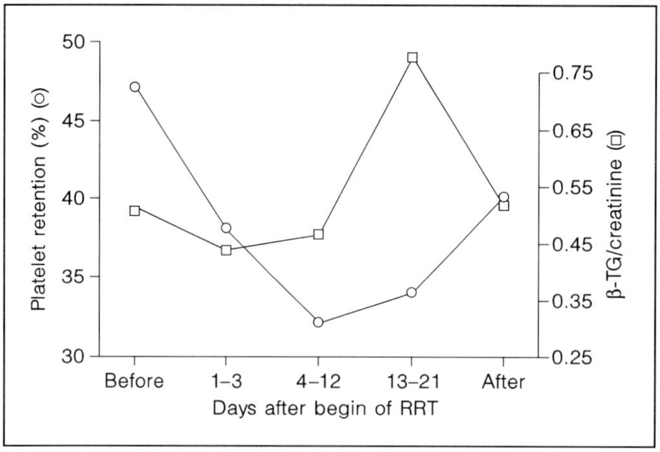

Fig. 2. Platelet activation markers (platelet retention; β-thromboglobulin/creatinine ratio) during renal replacement therapy (RRT) both in CVVH and HD patients (n = 35).

(HD 41 ± 30%, CVVH 31 ± 26%) and the β-thromboglobulin/creatinine ratio higher (HD 0.41 ± 0.29, CVVH 0.58 ± 0.43) (fig. 3, 4). The frequency of bleeding requiring transfusion was the same in both groups (HD 28%, CVVH 29%).

Discussion

There is a pronounced activation of coagulation with enhanced plasma coagulation marker levels (fibrinopeptide A, thrombin-antithrombin III complex) in patients with acute renal failure prior to renal replacement treatment. This hypercoagulability is due to the intravascular fibrin formation in the frequently coexisting DIC and apparently independent of blood contact with foreign surfaces. A similar hypercoagulability was described by Boisclair et al. [7] in patients with end-stage renal failure, irrespective of renal replacement therapy. During the course of treatment an additional activation of plasma coagulation and a difference between the two treatment groups cannot be observed by the markers in this study.

β-Thromboglobulin is catabolized by the kidney and its plasma concentration increases significantly when renal function is restricted. A significant positive correlation between serum creatinine and the β-thromboglobulin plasma concentration in patients with renal failure has been noted in a number of studies. Scharf

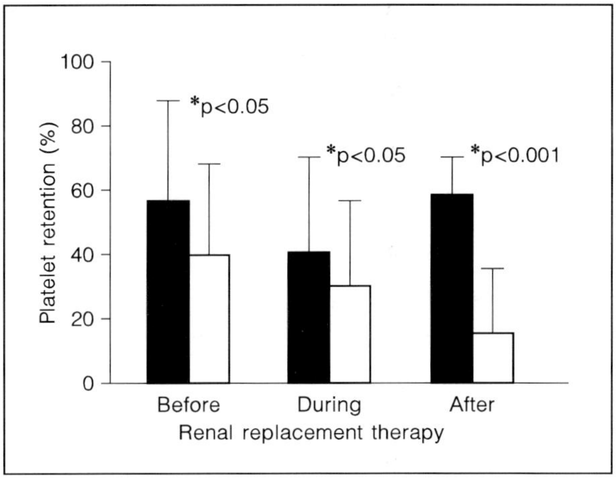

3

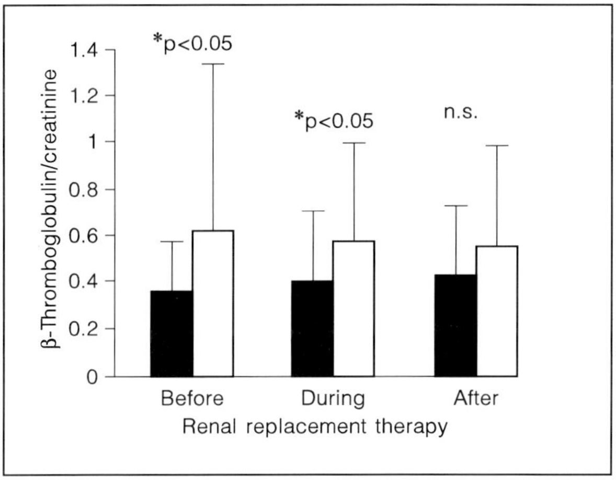

4

Fig. 3. Platelet retention in CVVH (□, n = 18) and intermittent HD (■, n = 17) treatment groups.

Fig. 4. β-Thromboglobulin/creatinine ratio in CVVH (□, n = 18) and intermittent HD (■, n = 17) treatment groups.

and Schneider [8] therefore suggested using the β-thromboglobulin/creatinine ratio to supplement platelet factor 4 as the most important marker of in vivo platelet activation during renal insufficiency. Platelet factor 4 increases significantly during heparin therapy, precluding its use with our patient population (heparinization) [9].

In acute renal failure, platelet dysfunction (thrombocytopathy), with decreased platelet retention, already existed before renal replacement commenced.

This may be explained most readily by the existing uremia. Renal replacement therapy eliminating uremic toxins failed however to reverse impairment of platelet function. Platelet retention decreased during renal replacement. Contact with foreign surfaces may cause an activation of platelets leading to adhesion on filter membranes and partial release of platelet contents (e.g. β-thromboglobulin) with subsequent slight temporary thrombocytopenia (platelet 'consumption') and defective platelet function [10]. During renal replacement therapy, platelet retention was lower and the β-thromboglobulin/creatinine ratio higher in the CVVH than in the HD group. The continuous contact with foreign surfaces in CVVH causes a more pronounced impairment of platelet function than the intermittent treatment.

References

1 Angelkort B: Thrombophilie und hämorrhagische Diathese bei chronischen Nierenkrankheiten. Nieren-Hochdruckkrankheiten 1979;1:36–43.

2 Lindsay RM, Prentice CR, Davidson JF, Burton JA, McNicol GP: Haemostatic changes during dialysis associated with thrombus formation on dialysis membranes. Br Med J 1972;iv:454–458.

3 Kramer P, Wigger W, Rieger J, Matthaei D, Scheler F: A new and simple method for treatment of overhydrated patients resistant to diuretics. Klin Wochenschr 1977;55:1121–1122.

4 Ludlam CA, Cash JD: Studies on the liberation of β-thromboglobulin from human platelets in vitro. Br J Haematol 1976;33:239–247.

5 Pelzer H, Schwartz A, Heimburger N: Determination of human thrombin-antithrombin III complex in plasma with an enzyme-linked immunosorbent assay. Thromb Haemost 1988;59:101–106.

6 Hellem AJ: Platelet adhesiveness in von Willebrand's disease. A study with a new modification of the glass bead filter method. Scand J Haematol 1970;7:374–382.

7 Boisclair M, Lane DA, Ireland H, Philippou H, Esnouf P: Proteinase cascade activation in human renal failure. Impaired proteinase regulation in haemodialysis as a mechanism for thrombogenesis. Thromb Haemost 1993;69:778.

8 Scharf RE, Schneider W: Bestimmung der Plättchensekretion und Thrombinaktivität bei Patienten mit chronischer Niereninsuffizienz. Lab Med 1984;7:358–364.

9 Dawes J, Smith RC, Pepper DS: The release, distribution, and clearance of human beta-thromboglobulin and platelet factor 4. Thromb Res 1978;12:851–861.

10 Notohamiprodjo M, Andrassy K, Bommer J, Ritz E: Dialysis membranes and coagulation system. Blood Purif 1986;4:130–141.

Norbert Maurin, MD, Medical Clinic II, University Hospital Aachen,
D–52057 Aachen (Germany)

Sieberth HG, Stummvoll HK, Kierdorf H (eds): Continuous Extracorporeal Treatment in Multiple Organ Dysfunction Syndrome. Contrib Nephrol. Basel, Karger, 1995, vol 116, pp 145–149

Influence of Coagulation Parameters on Filter Running Time during Continuous Venovenous Hemofiltration

Ioannis Stefanidis, Johannes Hägel, Norbert Maurin

Medical Clinic II (Director: Prof. *H.G. Sieberth*), University Hospital, Aachen, Germany

In 1977, Kramer et al. [1] reported on a continuous treatment method for patients with acute renal failure: continuous arteriovenous hemofiltration. Various other continuous treatment methods, like continuous venovenous hemofiltration (CVVH), have since been developed.

Thrombotic occlusion of the extracorporeal system in CVVH can be prevented by means of continuous intravenous anticoagulation with heparin, ensuring an adequate filter running time, which is essential for a successful CVVH treatment. A heparin-independent influence of the plasma proteins, especially fibrinogen, and the blood cells (hematocrit, platelets) on filter running time is often postulated in the literature [2, 3]. In this study we evaluated the effects of pre-existing hemostatic alterations on filter running time during CVVH.

Subjects and Methods

We investigated 60 patients with acute renal failure (40 men and 20 women; mean age 63 ± 13 years). Filter running time, heparin dose, antithrombin III (AT III), fibrinogen concentration and global coagulation parameters (activated clotting time, ACT; partial thromboplastin time, PTT; thrombin time) were registered daily.

A standardized CVVH technique was employed. A percutaneously introduced double-lumen venous catheter (Shaldon catheter, 1.8 mm diameter) was used for vascular access (internal jugular vein, subclavian vein, femoral vein), All filters used were polysulfone hemofilters with a filtration surface of 1.35 m^2 (Ultraflux AV 600® ; Fresenius, Oberursel, Germany). Blood flow was approximately 120 ml/min; substitution fluid was administered in a postdilution mode. Some 30 liters/day were exchanged to maintain urea concentration at about 20 mmol/l. An unfractionated heparin (Liquemin®; Roche, Basel, Switzerland) was used as anticoagulant. Heparin dose was controlled via the ACT (>110 s).

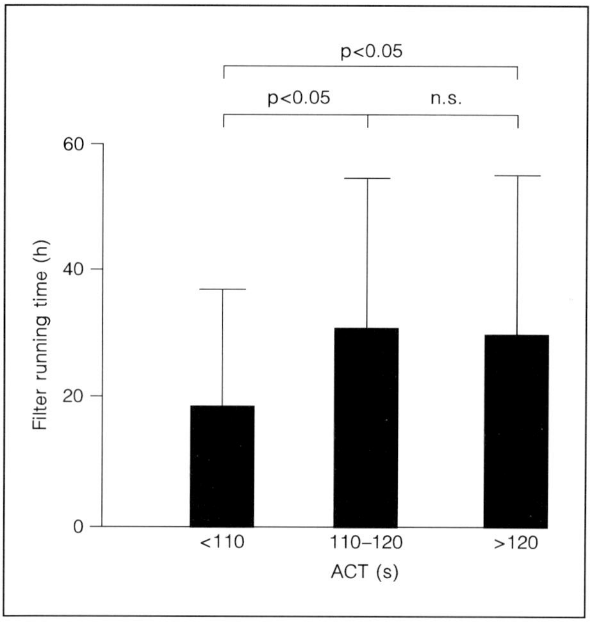

Fig. 1. Influence of ACT on filter running time during CVVH (n = 60).

Table 1. Correlation of hematological and hemostatic parameters with filter running time and heparin dose (n = 60)

Variable	Value (mean ± SD)	Correlation with[1]	
		filter running time	heparin dose
Heparin dose, IU/h	750 ± 467	–0.032	–
ACT, s	132 ± 22	0.001	–0.158*
PTT, s	54 ± 23	–0.034	–0.007
Thrombin time, s	52 ± 55	–0.048	0.434**
AT III, IU/ml	18 ± 5	0.007	0.268**
Fibrinogen, mg/dl	534 ± 186	–0.005	0.057
Platelet count, × 10^3/µl	151 ± 117	0.045	0.274**
Hematocrit, %	29 ± 3	–0.007	–0.041

* $p < 0.01$; ** $p < 0.001$.
[1] Correlation coefficients are given.

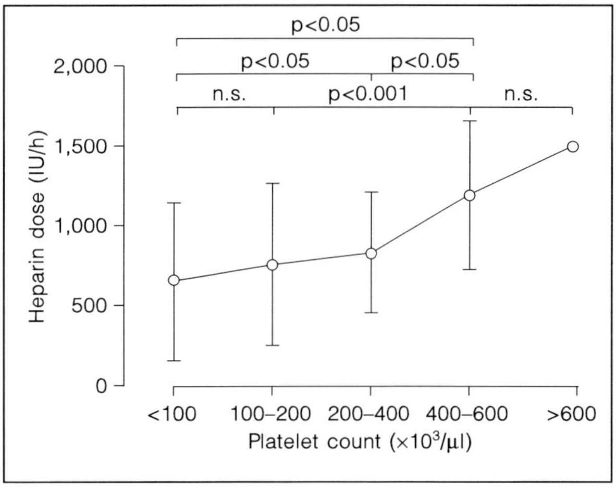

Fig. 2. Influence of increasing platelet count on heparin dose during CVVH (Pearson correlation coefficient = 0.274, n = 60, $p < 0.001$).

Results

The total treatment time for the entire patient population (n = 60) was 360 days, with a mean hemofiltration time of 6 days/patient. The mean filter running time was 30 ± 25 h; an average of 4–5 hemofilters were used for each patient. There were extreme inter- and intraindividual fluctuations in filter running time (variation: 1–125 h).

Filter running time correlated with ACT values < 120 s ($r = 0.2$, $p < 0.05$), but not with values >120 s (fig. 1). No correlation was observed between filter running time and the heparin dose, fibrinogen concentration, thrombin time or PTT (table 1). The fibrinogen concentration fluctuated over a wide range (80–1,300 mg/dl). Filter running times for patients with an extreme hyperfibrinogenemia (>600 mg/dl) did not differ significantly from those for the remaining patients (30 ± 26 h as opposed to 30 ± 23 h).

Correlation analysis revealed no significant influence of hematocrit or platelet count on filter running time. There was a significant correlation ($p < 0.05$) between the heparin dose (750 ± 467 IU/h) and ACT ($r = -0.158$) or thrombin time ($r = 0.434$), but not between the heparin dose and PTT ($r = -0.007$; table 1). In addition, there was a highly significant positive correlation between the platelet count and the required heparin dose ($r = 0.274$, $p < 0.001$; fig. 2.).

Discussion

A long filter running time (>24 h) is essential for an effective CVVH treatment. The permeability or loss of permeability of the hemofilters depend chiefly on their material and structural properties (effective membrane surface, pore size) and vary widely from type to type [4]. This factor played no role in our study, in which all hemofilters used were of the polysulfone type, with a filtration surface of 1.35 m^2. It is also suspected that an increase in fibrinogen concentration, platelets or hematocrit may shorten filter running time owing to more frequent thromboses in the extracorporeal system. In certain cases with extreme hyperfibrinogenemia (>600 mg/dl), heparinization was supplemented by defibrinogenation with ancrod (Arvin®), to improve extracorporeal flow [2, 3].

In our study, filter running time is dependent on ACT. There is, however, no significant correlation of filter running time with fibrinogen concentration, PTT, thrombin time, the heparin dose, platelet count or hematocrit. On the basis of these results, one must advise against general use of defibrinogenation for patients with high fibrinogen concentrations (>600 mg/dl) during CVVH, especially since administration of ancrod is regarded as obsolete due to the numerous associated complications [5].

The lack of correlation between filter running time and hematocrit is an unexpected result. Theoretically, hematocrit should strongly influence filter running time: firstly because of the reduced blood viscosity and secondly because of the diminished platelet adhesion associated with lowered hematocrit. The influence of increasing hematocrit on platelet adhesion, e.g. after erythropoietin therapy, has been demonstrated by in vitro and in vivo studies of patients on chronic hemodialysis [6, 7]. The interpretation of our results on the hematocrit independence of filter running time is, however, limited by the fact that 83% of all measured hematocrit values were <30%. Hematocrit fluctuations between 20 and 30% have no substantial influence on blood viscosity; an effect is not observed until values exceed 30% [8].

A platelet count >400,000/µl leads to an increase in the required heparin dose. This influence may most readily be attributed to the heparin-neutralizing effect of platelet factor 4 (a protein of the thrombocytic α-granules) [9].

References

1 Kramer P, Wigger W, Rieger J, Matthaei D, Scheler F: A new and simple method for treatment of overyhdrated patients resistant to diuretics. Klin Wochenschr 1977;55:1121–1122.

2 Maurin N: Antithrombotic management with a stable prostacyclin analogue during extracorporeal circulation. Contrib Nephrol. Basel Karger, 1991, vol 93, pp 205–209.

3 Roy J, Guidoin R, Martin L, Lephat H, Blais P, Marois M, Gagnon D, Awad J: Defibrinogenation as an alternative to heparinisation in prolonged extracorporeal circulation. Res Exp Med 1980;176: 219–234.

4 Jenkins RD, Kuhn RJ, Funk JE: Permeability decay in CAVH hemofilters. ASAIO Trans 1988;34: 590–593.
5 Latallo ZS: Retrospective study on complications and adverse effects of treatment with thrombin-like enzymes – A multicentre trial. Thromb Haemost 1983;50:604–609.
6 Bjornson J, Brostad F: Platelet and fibrin(ogen) deposition in the artificial kidney. The influence of haematocrit, fibrin monomer and platelet inhibitors. An in vitro study. Scand J Urol Nephrol 1978; 12:259–264.
7 Maurin N, Fitzner S, Fritz H, Gladziwa U, Hägel J, Stefanidis I: Influence of recombinant human erythropoietin on hematological and hemostatic parameters with special reference to microhemolysis. Clin Nephrol, in press.
8 Schmid-Schönbein H: Microrheology of erythrocytes, blood viscosity and the distribution of blood flow in the microcirculation. Int Rev Physiol 1976;9:1–6.
9 Lane DA, Peijler G, Flynn AM, Thompson EA, Lindahl UL: Neutralization of heparin-related saccharides by histidine-rich glycoprotein and platelet factor 4. J Biol Chem 1986;261:3980–3986.

Norbert Maurin, MD, Medical Clinic II, University Hospital Aachen,
D–52057 Aachen (Germany)

Sieberth HG, Stummvoll HK, Kierdorf H (eds): Continuous Extracorporeal Treatment in Multiple Organ Dysfunction Syndrome. Contrib Nephrol. Basel, Karger, 1995, vol 116, pp 150–153

The Mechanism of Postoperative Blood Loss Reduction by Ultrafiltration in Cardiac Surgery Is Unlikely to Be Only Mediated by Hemoconcentration of Coagulation Factors

Didier Journois[a], *Emmanuelle de Raucourt*[b], *Anne Marie Fischer*[b], *Guy Allanic*[c], *Moise Keita*[c], *Nicole de Courcy*[c]

Departments of
[a] Anesthesia and Intensive Care Medicine,
[b] Hematology, and
[c] Pediatric Cardiovascular Surgery, Laennec Hospital, Paris, France

The systemic inflammatory response that occurs in infants and children after cardiopulmonary bypass (CPB) results in a capillary leak syndrome that remains a major cause of morbidity and mortality [1]. In addition, the CPB-associated hemodilution of platelets and coagulation factors promotes the hemostatic impairment which is generally observed in these children [2]. Because the hemodilutional effects of CPB are especially pronounced in children due to a disproportionally large priming volume of the CPB circuit, the potential benefit of ultrafiltration could be very significant in these small patients. This technique has effectively been reported to be an effective therapy to reduce postoperative blood loss [3, 4]. These studies have shown that ultrafiltration is able to provide blood hemoconcentration at CPB termination. However, the mechanism of blood loss reduction has not been investigated and hemoconcentration of coagulation factors remains a speculative mechanism. The current study was intended to determine the effects of ultrafiltration on biological and clinical hemostasis variables.

Material and Methods

Thirty-two children undergoing surgical correction of tetralogy of Fallot were randomly assigned to be ultrafiltered during CPB rewarming or a control group. Cardiopulmonary bypass was performed with a stretching roller pump (RP06®; Rhône-Poulenc) and an appropriately sized Dideco hollow-fiber oxygenator was used (Dideco Lab., Mirandola, Italy). The tubing of the extracorporeal circuit was constructed of polyvinylchloride or silicone (Dideco). The extracorporeal circuit was primed with 35% volume of 40 g/l albumin solution (CNTS, Paris, France), 8% volume of molar sodium bicarbonate, 40% volume of hydroxyethyl starch (Elohes®; Biosedra Lab., Louviers, France) 5% aprotinin (10,000 IU/ml, Trasylol®; Bayer Lab., Puteaux, France), 12% of Hartmann's solution and fresh-frozen plasma or red blood cells according to patients' needs, to reach a total prime volume of 1,250 ml/m^2 with a minimum of 520 ml. Heparin was added to the priming solution (2 IU/ml). Anticoagulation was achieved with an initial bolus of heparin (beef lung sodium heparin; Léo Lab., Paris, France) of 250 IU/kg injected in the right atria before cannulation and followed by a continuous infusion of 62.5 IU/kg/h until the end of CPB. Aprotinin was administered in all children at a dose of 30,000 IU/kg after the induction of anesthesia and then continuously infused at a rate of 1.35 kIU/kg/min. After CPB, protamine (Choay Lab., Paris, France) was administered at a rate of 10 mg/min up to a total dose of 3.5 mg/kg. A cell separator system (Cell-Saver System IV, Haemonetics) was used in the two groups to wash and centrifuge blood aspirated in the surgical field prior to administration of heparin and as the sole aspiration method following the administration of protamine.

A Spiraflo polysulfone ultrafilter (HFT02®; Sorin Lab., Antony, France) was rinsed with 1,000 ml saline then inserted between the arterial tubing and the cardiotomy reservoir. Ultrafiltration was carried out with a rate adjusted to reach a cardiotomy reservoir level of zero at the completion of rewarming. Administration of red separator concentrate cells, albumin or fluids during rewarming was allowed excepting any solution containing procoagulant factors. After termination of CPB, the blood remaining in the CPB circuit was washed and centrifuged using the cell separator system and then re-transfused in the operating room or in the intensive care unit (ICU).

Arterial blood sampling was obtained during the aortic cannulation (T_1), prior to rewarming (and ultrafiltration) (T_2), at rewarming (and ultrafiltration) completion (T_3) and 24 h later in the ICU. The blood levels analyzed at these four different times were the following: plasma protein concentration and hemostatic variables including platelet count, coagulation factors II, V, VII + X, fibrinogen, Celite-activated clotting time (ACT), prothrombin time, antithrombin-III (AT) and D-dimers. The ICU physicians managing the patients were blinded to group assignment. Data were analyzed using the Mann-Whitney nonparametric method for unpaired data. Statistical significance was chosen at $p < 0.05$.

Results

There were no significant differences between the ultrafiltration and the control groups with respect to duration of bypass, aortic cross-clamping duration (51.5 ± 15 vs. 56.4 ± 23 min), age (39 ± 30 vs. 55 ± 43 months) (fig. 1). Cumulative blood loss was significantly lower in the ultrafiltered group compared to the control group at the 24th hour (respectively 250 [176–356] vs. 319 [182–500] ml/m^2). Significant, but mild increases in clotting factors, AT-III and hematocrit (Hct) were noted for the ultrafiltration group: FII 7.5 ± 7%, FV 5.2 ± 7%, FVII+X 5 ± 8%, AT-III 7 ± 6%, Hct 10 ± 18%, proteins 10.5 ± 18 g/l. These

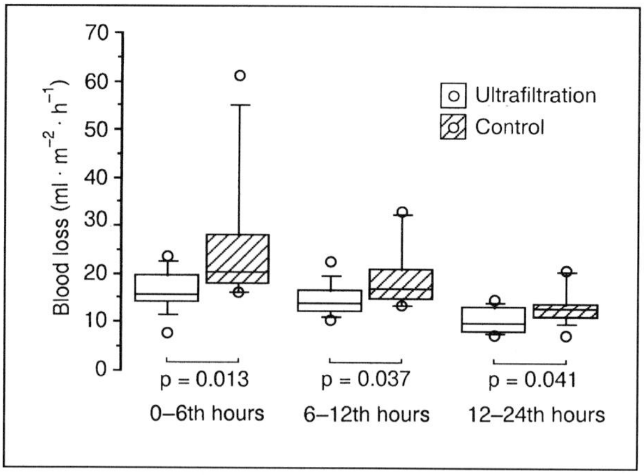

Fig. 1. Differences in postoperative blood loss between ultrafiltration and control groups. Mann-Whitney nonparametric test for unpaired data. Data are expressed with box plots. The lower boundary of the box is the 25th percentile, and the upper is the 75th percentile. The median value is in the box. Lines are drawn from the end of the box to the largest and smallest observed values that are not outliers.

changes were the following in the control group: FII –3 ± 3%, FV 0.25 ± 2.2%, FVII+X –0.17 ± 1.5%, AT-III –1.2 ± 4.4%, Hct 5.5 ± 5%, proteins 5.5 ± 5 g/l. No difference was observed in platelets count and D-dimers. The amount of administered fluid loading during the warming phase of CPB was 554 ± 372 ml/m^2 in the hemofiltration group vs. 500 ± 307 ml/m^2 in the control group respectively composed of 70 vs. 61% of red separator concentrate cells (p = 0.62), 20 vs. 24% of exogenous blood cells (p = 0.55) and 10 vs. 14% of 4% albumin (p = 0.12). There were no reoperations for bleeding. No donor blood was transfused postoperatively in either group. No difference was observed between groups at the 24th hour in ICU regarding hemostasis parameters.

Discussion

The postoperative bleeding reduction produced by ultrafiltration has already been observed by Naik et al. [4]. They speculated that clotting factors and platelets are concentrated by ultrafiltration leading to improved clotting conditions [4]. Despite a significant blood loss reduction, our study shows that ultrafiltration only provides a limited increase in coagulation factor concentration and no change in platelet count. These results suggest that, when ultrafiltration is performed during CPB rewarming, its relevant effects on postoperative bleeding are unlikely to be solely explained by hemoconcentration. The ability of ultrafiltration to remove some major mediators of the inflammatory response could be an explanation for this phenomenon, and needs to be investigated [5].

References

1 Kirklin JK, Blackstone EH, Kirklin JW: Cardiopulmonary bypass: Studies on its damaging effects. Blood Purif 1987;5:168–178.

2 Kern FH, Morana NJ, Sears JJ, Hickey PR: Coagulation defects in neonates during cardiopulmonary bypass. Ann Thorac Surg 1992;54:541–546.

3 Elliott MJ: Ultrafiltration and modified ultrafiltration in pediatric open heart operations. Ann Thorac Surg 1993;56:1518–1522.

4 Naik SK, Knight A, Elliott M: A prospective randomized study of a modified technique of ultrafiltration during pediatric open-heart surgery. Circulation 1991;84(suppl 5):III422–431.

5 Millar AB, Armstrong L, van der Linden J, Moat N, Ekroth R, Westwick J, Scallan M, Lincoln C: Cytokine production and hemofiltration in children undergoing cardiopulmonary bypass. Ann Thorac Surg 1993;56:1499–1502.

Dr. D. Journois, Hôpital Laennec, 42, rue de Sèvres, F–75007 Paris (France)

Sieberth HG, Stummvoll HK, Kierdorf H (eds): Continuous Extracorporeal Treatment in Multiple Organ Dysfunction Syndrome. Contrib Nephrol. Basel, Karger, 1995, vol 116, pp 154–158

Antithrombin III Deficiency during Continuous Venovenous Hemodialysis

O. Bastien, P. French, S. Paulus, S. Filley, M. Berruyer, M. Dechavanne, S. Estanove

Intensive Care Department, Hôpital Cardiovasculaire et Pneumologique Louis-Pradel, Lyon, France

Postoperative acute renal failure (ARF) is often multifactorial [1, 2] and associated with multiorgan dysfunction syndrome (MODS). For these reasons, concomitant liver failure with disseminated intravascular coagulopathy (DIC) could be associated with renal failure. Anticoagulation is necessary to perform a hemodialysis procedure. Data concerning acquired deficiency of antithrombin III (ATIII) are scarce [3, 14] in this case, specially if heparin should be used. The aim of the study is to evaluate the frequency and risk factors of ATIII deficiency in ARF.

Methods

Fifty consecutive patients scheduled for cardiac surgery were studied after postoperative ARF. No patients were excluded. The first measurement of ATIII was performed during the first day of hemodialysis. ATIII as a heparin cofactor was measured by an amidolytic technique. The hemodialysis technique was a continuous venovenous hemodiafiltration using PAN 69 membrane and a blood pump (BSM 22, Hospal).

Anticoagulation used a continuous infusion of heparin in the circuit, in order to obtain a titration of heparin >0.30 IU/ml. Twice-daily coagulation tests were obtained to detect DIC and control the blood heparin level. Acquired ATIII deficiency was defined by a limit level of 60%, as generally accepted [3, 8, 19]. For severity scoring of the patients we used the Apache II score [2] and MODS definitions of Bone [1]. Liver failure was defined by a total bilirubin up to 76 mmol/l.

Statistics included Anova and Student's test for comparison of means between groups with Bonferroni correction for repeated tests, and stepwise analysis of risk factors.

Table 1. Comparison of data between groups with or without ATIII deficiency

	ATIII deficiency group	Control group
Apache II	24.9±6	21.7±4*
MODS score	2.9±0.8	2.2±0.8*
Male gender, %	95	65
Age, years	52.3±12	60.5±11*
Mortality, %	89	39
Prothromobin time, % normal	36.7±17	63.3±12*
Fibrinogen, g/l	2.8±1.4	5.0±1.3*
Platelets, 10^9/l	87.9±37	99.6±83
APTT, s	87.7±37	80.6±47
Bilirubin, μmol/l	263±184	82±118*
ICU stay, days	17.5±17	29.6±23
Dialysis duration, days	8.3±7	6.9±4
Membrane duration, days	1.7±0.7	2.3±0.7*

* $p < 0.01$.

Results

An acquired ATIII deficiency (AAD) was found in 12 patients (38%) with a mean level of 37.3 ± 11% (AAD group) vs. 79.3 ± 14% in the control group. The two groups were not significantly different for inotropic support and cardiac pathology (valvular 20%, transplantation 34%, vascular 20%, ischemic and miscellaneous 26%. The two groups were different (table 1) for Apache score, MODS, age and sex ratio. The overall mortality was double for the AAD group. Prothrombin time and fibrinogen were lower in the AAD group, but did not differ for platelet counts and APTT. Liver failure was more frequent in the AAD group (84 vs. 23%), with a greater cholestasis. Sepsis was not significantly different (78 vs. 68%).

Liver failure was statistically associated with male gender, but without any difference concerning the use of steroids. Adverse events or clinical management data (table 2) differ for neurological disorders (50 vs. 29%). The mean duration of the hemodialysis membrane was significantly shorter in the AAD group (1.7 ± 0.7 vs. 2.3 ± 0.7 days). Changing the membrane was performed after thrombosis, decrease of hourly ultrafiltration rate <100 ml, or systematically after 72 h. The

Table 2. Stepwise regression analysis

	Variable		
	regression weights beta	multiple R	F if significant
Fibrinogen	0.461	0.693	20.4
Neurological events	0.442	0.748	3.75
Sex	0.462	0.806	5.11
Hemorrhage	-0.260		
Sepsis	0.313		
Steroids	-0.139		
MODS	0.195		
Mechanical ventilation	0.126		
Cardiac etiology	0.248		

worst prognosis of the AAD group was not associated in stepwise analysis to hemorrhage (28%), higher duration of ICU study, hemodialysis period (8.3 ± 7 vs. 6.9 ± 4 days), or mechanical ventilation.

Discussion

The diagnosis of ATIII deficiency may be difficult using standard monitoring of anticoagulation. APTT in case of complex liver dysfunction of inflammatory disease (with factor XII abnormality) does not reflect the activity of the complex heparin-ATIII. On the other hand, direct determination of blood heparin level is performed with standardized ATIII level, so that the deficiency could be difficult or forgotten.

In anuric patients, AAD has been described in chronic renal failure [9], specially in nephrotic syndrome [10, 11]. Dialysis with heparin does not seem to influence the risk [12, 13]. Sepsis and DIC have been reported to be the more important factors [3, 14]. The level of ATIII in 25% [14] of these patients was <55%, and similar to our results.

The mechanism of this disorder has been correlated to a decrease of its half-life [4, 5], sometimes to a direct alteration of hepatic synthesis. Another mechanism emphasizes an increase of ATIII clearance [15] mediated by the hepatic receptors (Serpin Receptor 1). Sepsis could be a cofactor inducing DIC, liver dysfunction, but also leukocyte stimulation. This could be responsible for elastase release [16, 17], which could inhibit ATIII receptors on the endothelial sites. This inhibitory effect could be increased by heparin [18]. Sepsis and AAD could have an important link [5, 16, 19] and a high mortality.

The risk of AAD in ARF seems to be real when observing mortality in this group. The interest of a substitution trial has to be confirmed [19, 20]. Nevertheless, a standard protocol of heparin could not be proposed in these patients, specially in cardiac patients where anticoagulation is useful not only for the dialysis procedure but also for cardiac reasons. Other protocols like low-weight heparin and citrate also have to be discussed.

Conclusion

Monitoring ATIII seems to be interesting in ARF, as a prognostic index, and to assure the efficiency of heparin anticoagulation, if this procedure is used during dialysis, specially in cardiac patients.

References

1 Consensus Conference: Definitions for sepsis and organ failure and guidelines for the use of innovative therapies in sepsis. Crit Care Med 1992;20:864–874.

2 Knaus W, Draper E, Wagner D, Zimmerman J: Apache II: A severity of disease classification system. Crit Care Med 1985;13:818–829.

3 Vinazzer H: Therapeutic use of antithrombin III in shock and disseminated intravascular coagulation. Semin Thromb Hemost 1989;15:347–352.

4 Blauhut B, Necek S, Kramar H, Vinazzer H: Activity of antithrombin III and heparin on coagulation in shock. Thromb Res 1980;19:775–782.

5 Fourrier F, Lestarel P, Chapin C, Marey A, Goudemaud J: Traitement des coagulations intravasculaires disséminées septiques par l'antithrombine III. Réanim Soins Intens Med Urg 1987;2:281.

6 Fourrier F: Traitement des coagulations intravasculaires disséminées septiques: Héparine ou antithrombine III? Réanim Soins Intens Med Urg 1991;7:365–369.

7 Maki M, Terao T, Ikenoue T, Takemura T, Sekiba K, Shirakawa K, Soma H: Clinical evaluation of antithrombin III concentrate for disseminated intravascular coagulation in obstetrics. Gynecol Obstet Invest 1987;23:230–240.

8 Blauhut B, Kramar H, Vinazzer H: Substitution of antithrombin III in shock and DIC: A randomized study. Thromb Res 1985;39:81–89.

9 Toulon P, Jacquot C, Capron L, Frydman MO, Vigon D, Aiach M: Antithrombin III and hepatic cofactor II in patients with chronic renal failure undergoing regular hemodialysis. Thromb Haemost 1987;57:263–268.

10 Thaler E, Balzar E, Kopsie H, Pinggera WF: Acquired antithrombin III deficiency in patients with glomerular proteinuria. Haemostasis 1978;7:257–272.

11 Kauffmann RH, Keltkamp JJ, Tilburg NH, Van ES: Acquired antithrombin III deficiency and thrombosis in the nephrotic syndrome. Am J Med 1978;65:607–613.

12 Jorgensen M, Eriksen HO, Tranebjaerg L: Plasma antithrombin III concentration in patients on regular haemodialyis treatment. Nephron 1985;40:22–24.

13 Woo KT, Wei SS, Lee EJ, Lau YK, Lim CH: Effects of hemodialysis and peritoneal dialysis on antithrombin III and platelets. Nephron 1985;40:25–28.

14 Schuster HP, Konz MW, Blair J, Sedensky JA, Mammen EF: The influence of disseminated intravascular coagulation on renal function after experimental hemorrhagic shock. Resuscitation 1980; 8:3–28.
15 Pizzo S: Serpin Receptor 1: A hepatic receptor that mediates the clearance of antithrombin III proteinase complexes. Am Med 1989;87:3, 10–14.
16 Fourrier F, Chopin C, Goudemand J, Hendrycx S, Caron C, Rime A, Marey A, Lestavel P: Septic shock, multiple organ failure and disseminated intravascular coagulation. Chest 1992;101:816–823.
17 Bone R: Sepsis and coagulation, an important link. Chest 1992;101:594–595.
18 Jordan R, Kilpatrick J, Nelson R: Heparin promotes the inactivation of antithrombin by neutrophil elastase. Science 1987;237:777–779.
19 Büller H, Cate JW: Acquired antithrombin III deficiency: Laboratory diagnosis, incidence, clinical implications and treatment with antithrombin III concentrate. Am J Med 1989;87:3B, 44–48.
20 Emerson Th, Fournel M, Redens Th, Taylor F: Efficacy of antithrombin III supplementation in animal models of fulminant *E. coli* endotoxemia or bacteremia. Am J Med 1989;87:3B, 27–32.

Dr. O. Bastien, Anesthesia and Intensive Care Department, Hôpital Cardiologique,
boulevard Pinel, F–69394 Lyon Cedex 03 (France)

Sieberth HG, Stummvoll HK, Kierdorf H (eds): Continuous Extracorporeal Treatment in Multiple Organ Dysfunction Syndrome. Contrib Nephrol. Basel, Karger, 1995, vol 116, pp 159–162

Thromboelastography and Monitoring of Coagulation in Patients Undergoing Continuous Venovenous Hemofiltration

M. Derrier[a], *P. Jambou*[a], *M. Kaidomar*[a], *C. Ichai*[a], *J. Bayle*[b], *D. Grimaud*[a]

[a] Department of Anesthesiology and Intensive Care, and
[b] Laboratory of Hematology, University Hospital, Nice, France

Continuous hemofiltration (CHF), widely used as a renal replacement procedure in intensive care units, requires efficient and continuous antithrombotic therapy to prevent filter clotting without bleeding in patients [1–3]. In that case, low weight molecular heparin (LWMH) appears more convenient compared to standard heparin as anticoagulant, because of its superiority in reducing hemorrhagic tendency [4, 5]. In spite of the evident need to control coagulation, hemostasis tests able to detect risks of filter clotting or patient bleeding still remain to be identified. Thromboelastography (TEG), developed by Harttert in 1948, is based on measurement of in vitro whole blood viscoelastic properties and provides information on all phases of blood coagulation. It documents the interaction between platelets and coagulation proteins, from the initial time of fibrin-platelet interaction to eventual fibrinolysis. Easily performed, TEG has been viewed as a useful monitoring method for high bleeding risk surgery such as liver transplantation or cardiac surgery [6, 7]. The aim of this work was therefore to assess the ability of bedside TEG to detect risks of hemofilter clotting and bleeding in patients undergoing CHF.

Material and Methods

Twenty-two critically ill patients with acute failure (ARF), undergoing continuous venovenous hemodiafiltration (CVVHDF), were consecutively included in this study between March and July 1993. After femoral cannulation of an 11-French double lumen hemodialysis catheter, CVVHDF was performed for all patients via a venovenous circuit using a double-head pump monitoring module (BSM 22/VPM®; Hospal, Basel, Switzerland), and a 0.43 m^2 flat plate polyacrylonitrile hemofilter (AN-69S®; Hospal, France) changed every 48 h. Blood line thrombosis was prevented using a continuous infusion of sodic dalteparin

Table 1. Sensitivity and specificity of coagulation tests for detection of filter clotting

Coagulation test	Threshold value	Sensitivity	Specificity	Youden index
Hemochron, s	175	0	100	0
	450	98.55	1.43	0.02
Fibrinogen, $g \cdot l^{-1}$	1	22.86	76.92	0.22
	7	100	0	0
Platelets, $G \cdot l^{-1}$	10	2.38	96.3	1.32
	120	95.24	3.7	1.06
TEG r+k, mm	15	0	100	0
	60	90.16	11.29	-1.45
TEG MA, mm	20	0	100	0
	62.5	100	0	0

(Fragmin®) in doses adjusted to obtain anti-Xa levels between 0.1 and 0.5 $IU \cdot ml^{-1}$. Blood flow was kept at a constant rate of 100 ml/min and bicarbonate dialysate (Biosedra®, France) was delivered by volumetric countercurrent pump at a rate of 1 liter/h. Replacement fluid was administered in a postdilution mode according to the cardiovascular state of the patients and a central venous pressure monitoring. The ultrafiltrate was collected in a volumetric bag and measured hourly. Activated clotting time (Hemochron®) and whole blood TEG were used, every 8 h at bedside, to monitor coagulation while morning samples for laboratory tests (platelet count, fibrinogen level, prothrombin time (PT), partial thromboplastin time (PTT) and anti-Xa level (AXa)) were driven. Hemofilter clotting was detected clinically by ultrafiltration rate (UFR) $< 300\ ml \cdot h^{-1}$, and the need for changing filter, and biologically by urea extraction coefficient (UEC) < 0.25. Bleeding in patients was assessed by the number of hemorrhagic sites and blood transfusion requirements. Statistical analysis used two methods [8]: sensitivity and specificity studies for each coagulation parameter to detect hemofilter clotting, and correlation studies between coagulation parameters and clotting or bleeding criteria.

Results

Filter clotting, requiring replacement, was observed in 31% of filters in spite of patients' relative hypocoagulability profiles. Fifty percent of less than 48-hour-old filters had UEC values lower than 0.25. No significant difference was noted in coagulation profiles for detecting subclinical thrombosis between clotted and efficient filters. For each coagulation test, a good sensitivity was associated with no specificity (table 1), and no correlation was found between coagulation profiles and UFR (table 2). For bleeding detection, only a poor correlation was found between two TEG parameters (Emx, IPT) and number of hemorrhagic sites (table 3).

Table 2. Correlations between coagulation parameters and ultrafiltration rate

Coagulation test	R	p
Hemochron	0.06	0.9143
Fibrinogen	0.008	0.9005
Platelets	0.008	0.9005
TEG r+k	0.1	0.1354
TEG MA	0.12	0.07
TEG IPT	0.104	0.1211

Table 3. Correlations between coagulation parameters and number of hemorrhagic sites

Coagulation test	R	p
Hemochron	0.045	0.4149
Fibrinogen	0.104	0.1485
Platelets	0.328	0.0001
TEG r+k	0.123	0.0532
TEG MA	0.09	0.1581
TEG Emx	0.155	0.0142
TEG IPT	0.132	0.038

Discussion

Hypocoagulability profiles were found in patients with surgical diseases (polytraumas, major vascular surgery, liver transplantation, severe postpartum hemorrhage, postoperative hemorrhagic shock) or severe sepsis. In these patients, choice of a LWMH (Fragmin®) as antithrombotic method for CVVHDF was justified. However, clotting occurred in 31% of the hemofilters. If Fragmin® doses, adjusted to obtain AXa levels from 0.1 to 0.5 $IU \cdot ml^{-1}$, avoided iatrogenic hemorrhagic risk, they were probably insufficient to prevent all filter clotting. TEG and other coagulation tests failed to detect risk of thrombosis. This could be explained by two theories: (1) when using LWMH, no test allows measurement of antithrombotic activity and only hemorrhagic risks can be detected by AXa levels and chronometric tests, and (2) when patients undergo CHF, contact of blood with foreign surfaces is followed by adsorption and activation of platelets and coagulation proteins [2, 5]. Concentration of coagulation factors in filters is increased by plasma water removal, favoring local activation of coagulation cascade [2]. Clot activation in hemofilters seems to be independent of patient's coagulation status and the specific indicator of early filter clotting remains to be identified. Polymorphonuclear elastase level appears to be an indicator of coagulation activation [5], but this test is not yet available. So, detection of early filter clotting is based on both clinical criteria: decrease of UFR, membrane permeability index [9], or biological criteria: urea clearance, UEC.

Evaluation of bleeding is based on clinical criteria: blood losses in drains, transfusion requirements, number of hemorrhagic sites, and on biological criteria: red cell and platelet counts, coagulation tests. Whole blood TEG allows identification of coagulation disorders and structural data (MA, Emx, IPT) documents clot stability, which depends on the platelet-fibrinogen interaction [6, 10]. This interaction is not evaluated by common coagulation tests performed on plasma. For Spiess et al. [7], MA is correlated with platelet count in patients with bleeding after cardiac surgery. In our study, platelet count and MA values were correlated with number of hemorrhagic sites. However, this correlation was poor, and a dynamic study over time should prove useful to confirm these results.

Conclusion

When using LMWH (Fragmin®) for CHF, clotting activation in hemofilters appears to be independent of patient's coagulation status, and neither TEG nor other coagulation tests are able to detect early hemofilter thrombosis. Filter clotting must be detected only by using clinical (UFR) or biological (UEC) criteria. For risks of hemorrhages, structural abnormalities of TEG (Emx, IPT) are poorly correlated with number of bleeding sites and diffuse hemorrhages.

References

1 Jambou P, Ichai C, Grimaud D: L'hémofiltration continue: une méthode adaptée au traitement de l'insuffisance rénale aiguë en réanimation; in Mises Au Point en Anesthésie-Réanimation, Paris 1993, pp 677–710.

2 Journois D, Mauriat P: Prévention antithrombotique au cours de l'hémofiltration continue; in Journois D (ed): Hémofiltration continue. Paris, Elsevier, 1993, pp 63–78.

3 Mehta RL, Dobos GJ, Ward DM: Anticoagulation in continuous renal replacement procedures. Semin Dial 1992;5:61–68.

4 Schrader J, Stibbe W, Armstrong VW, Kandt M, Muche R, Köstering H, Seidel D, Scheler F: Comparison of low molecular weight heparin to standard heparin in hemodialysis/hemofiltration. Kidney Int 1988;33:890–896.

5 Swars H, Hafner G, Weilemann LSL, Ehrenthal W, Schinzel H, Prellwitz W, Meyer J: Acute dialysis: PMN elastase as a new parameter for controlling individual anticoagulation with low molecular weight heparin (Fragmin®). Intensive Care Med 1991;17:52–56.

6 Mallet SV, Cox DJA: Thromboelastography. Br J Anaesth 1992;69:307–313.

7 Spiess BD, Tuman KJ, McCarthy RJ, deLaria GA, Schillo R, Ivankovitch AD: Thromboelastography as an indicator of post-cardiopulmonary bypass coagulopathies. J Clin Monit 1987;3:25–30.

8 Rumeau-Rouquette C, Breart G, Padieu R: Analyse des études d'évaluation; in Méthodes en épidémiologie. Paris, Flammarion, 1985, pp 376–388.

9 Journois D, Safran D, Castelain MH, Chanu D, Drevillon C, Barrier G: Comparaison des effets antithrombotiques de l'héparine, l'énoxaparine et la prostacycline au cours de l'hémofiltration continue. Ann Fr Anesth Réanim 1990;9:331–337.

10 Dorman BH, Spinale FG, Bailey MK, Kratz JM, Roy RC: Identification of patients at risk for excessive blood loss during coronary artery bypass surgery: Thromboelastograph versus coagulation screen. Anesth Analg 1993;76:694–700.

M. Derrier, Department of Anesthesiology and Intensive Care, University Hospital, F–06000 Nice (France)

Sieberth HG, Stummvoll HK, Kierdorf H (eds): Continuous Extracorporeal Treatment in Multiple Organ Dysfunction Syndrome. Contrib Nephrol. Basel, Karger, 1995, vol 116, pp 163–168

Prognosis in Pediatric Patients with Multiple Organ System Failure and Continuous Extracorporeal Renal Support

G. Zobel, E. Ring, S. Rödl
Department of Pediatrics, University of Graz, Austria

Since acute renal failure (ARF) is often a part of multiple organ system failure (MOSF) in critically ill patients, the mortality rate in adults remains above 50% [1]. Over the past several years there has been a growing consensus that a form of continuous renal replacement therapy is the method of choice for renal replacement in critically ill patients with ARF [2–5]. The hallmark of these renal replacement therapies is their hemodynamic stability. The aim of this study is to look for risk factors in pediatric patients with MOSF and continuous extracorporeal renal support (CERS).

Patients and Methods

From June 1985 to June 1994, 84 critically ill infants and children with a mean age of 3.4 ± 0.5 years (SEM) and a mean body weight of 14.8 ± 1.9 kg (SEM) underwent continous arteriovenous or venovenous renal support at the pediatric ICU of the Children's Hospital, University of Graz. The clinical features of the patients are presented in table 1. Severity of illness was assessed daily during the first 4 days of CERS using the Acute Physiologic Score for Children (APSC), the Pediatric Risk of Mortality Score (PRISM) and the Organ System Failure Score (OSF) [6–8]. The amount of monitoring and therapeutic support was assessed by the Therapeutic Intervention Scoring System (TISS) [9].

During CAVH the blood lines were shortened to minimize the extracorporeal volume and to reduce the resistance of the blood lines. A roller pump (Gambro AK10™; Gambro Corp., Hechingen, FRG) with exchangeable pump headings was used for CVVH. The blood flow of the pump ranged from 12 to 75 ml/min and from 50 to 300 ml/min, respectively.

Table 1. Clinical features of critically ill infants and children with continuous renal replacement therapy (n = 84, mean ± SEM)

	Survivors (n = 46)	Nonsurvivors (n = 38)
Age, years	2.2 ± 0.5	4.8 ± 0.9**
Body weight, kg	10.7 ± 1.8	19.5 ± 3.4*
Sex ratio (M/F)	29/17	28/10
Diagnosis		
ARF		
Simple	2	0
Complicated	8	0
Low cardiac output	15	10
MOSF	11	27
Hypervolemia	4	0
Metabolic crisis	6	1
Urine flow (ml/kg/h)		
Oliguria (<0.5)	23	7
Anuria (<0.25)	19	30
Polyuria (>4.0)	4	1
Patients on		
Mechanical ventilation	40	38
Vasopressor support	34	37
Mean arterial pressure, mm Hg	63.7 ± 3.2	49.0 ± 2.0***
Mode of renal support		
Arteriovenous	26	15
Venovenous	20	23
Duration of renal support, h	149 ± 22.8	118.2 ± 22.1

* $p < 0.025$; ** $p < 0.01$; *** $p < 0.001$.

Vascular access for CVVH was obtained using 5.5- to 9-French double-lumen catheters (Medcomp™, Medical Components Inc., Harleysville, Pa., USA) inserted percutaneously using the Seldinger technique. Anticoagulation for both treatment modes was achieved with heparin (10 IU/kg/h) given continuously into the arterial line of the extracorporeal circuit. The ultrafiltrate was partially or totally replaced in the postdilution mode according to the clinical situation. The isotonic substitution fluid was either based on lactate or bicarbonate. Fluid in- and output was balanced by a microprocessor-controlled fluid control system (Equaline™; Amicon Corp., Lexington, Mass., USA). Data are given as mean ± SEM. Student's t test for unpaired samples (two-way analysis) and χ^2 test were used for comparing data of survivors and nonsurvivors.

Table 2. Pretreatment biochemical data of patients with continuous renal replacement therapy (n = 84)

	Survivors (n = 46)	Nonsurvivors (n = 38)
pH	7.29 ± 0.02	7.20 ± 0.03*
HCO_3, mmol/l	17.0 ± 1.0	16.4 ± 1.6
Lactate, mg/dl	80 ± 19	106 ± 44
SvO_2, %	66 ± 9.6	63 ± 3.0
Serum urea, mg/dl	118 ± 11.7	112 ± 11.9
Serum creatinine, mg/dl	2.5 ± 0.3	2.48 ± 0.3

SvO_2 = Mixed venous oxygen saturation.
* $p < 0.05$.

Table 3. Outcome and age of the patients with continuous renal replacement therapy (n = 84)

Age	Survivors	Nonsurvivors	Mortality, %
<1 month	19	10	34
1–12 months	6	6	50
1–5 years	15	9	37
>5 years	6	13	68
Total	46	38	45

Results

Mean duration of CERS was 125 ± 22 h. Biochemical data before starting CERS are given in table 2. The mean prehemofiltration serum creatinine levels of survivors and nonsurvivors were 2.5 ± 0.3 and 2.48 ± 0.3 mg/dl, the posthemofiltration levels were 1.9 ± 0.2 and 2.1 ± 0.2 mg/dl, respectively. The serum urea levels of survivors and nonsurvivors decreased from 118 ± 11.7 to 101 ± 7.5 mg/dl and from 112 ± 11.9 to 100 ± 7.1 mg/dl, respectively.

The survival rate of the total group was 55%. The survival rates for patients with ischemic and nephrotoxic/septic acute tubulus necrosis were 51.1 and 38.1%, respectively. Mortality rate in the CVVH group (53%) was higher than in

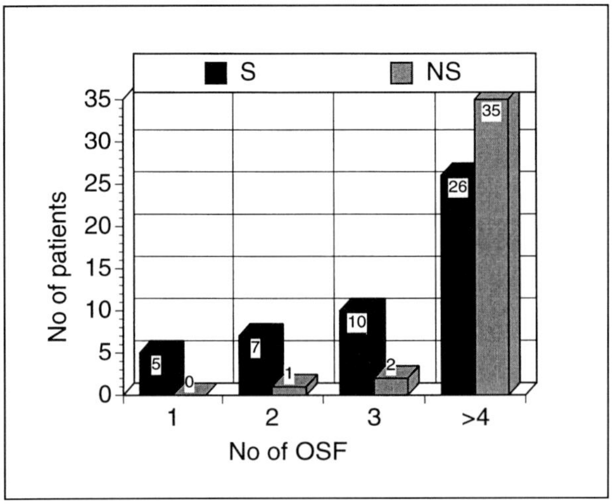

Fig. 1. Mortality rate and number of organ system failures (OSF) in critically ill infants and children with continuous renal replacement therapy. S = Survivors, NS = nonsurvivors.

Table 4. Scoring systems in patients with continuous renal replacement therapy (n = 84)

Scoring system	Survivors	Nonsurvivors
Acute Physiologic Score for Children (APSC)	20.1 ± 1.2	27.6 ± 1.4***
Pediatric Risk of Mortality (PRISM)	13.3 ± 1.0	21.9 ± 1.8***
Organ System Failure Score (OSF)	3.2 ± 0.2	4.2 ± 0.2***
Therapeutic Intervention Scoring System (TISS)	43.6 ± 1.8	48.5 ± 2.2

*** $p < 0.0001$.

the CAVH group (36%). As shown in table 3, the lowest mortality rate was observed in neonates (34%) and the highest in children older than 5 years (68%). Figure 1 presents the mortality rate according to the number of organ system failures. All patients with uncomplicated ARF survived. In patients with ARF and 1, 2 and >3 organ system failures, the mortality rate was 12.5, 16.6 and 57%, respectively.

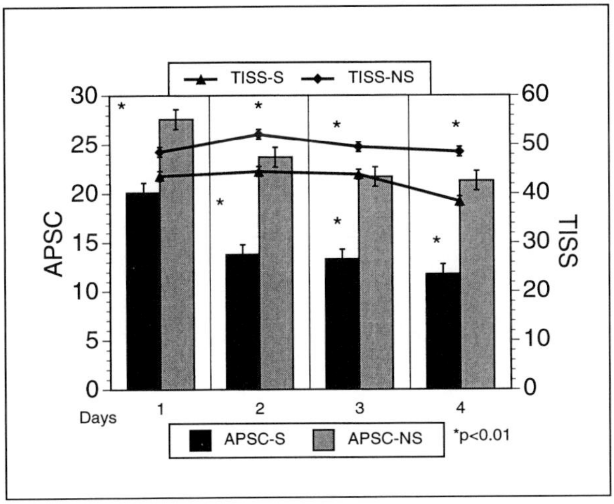

Fig. 2. Clinical scores in critically ill infants and children with continuous renal replacement therapy. S = Survivors, NS = nonsurvivors, APSC = Acute Physiologic Scores for Children, TISS = Therapeutic Intervention Scoring System.

There were significant differences in the severity of illness between survivors and nonsurvivors assessed by the APSC, PRISM and OSF scores (table 4). As shown in figure 2, the amount of monitoring and therapeutic support was not different between both groups within the first day of treatment. Whereas TISS points for survivors decreased significantly from 43.6 ± 1.8 to 37.4 ± 1.4 ($p < 0.05$) within the first 4 days of treatment, there was no change for nonsurvivors (48.5 ± 2.2 and 48.5 ± 0.3).

Discussion

The pediatric literature reports a mortality rate up to 83% in patients with ARF complicating medical or surgical illness, whereas ARF due to primary renal disease had only a 10% mortality [10]. The mortality rate in adults with uncomplicated ARF is below 20%. Once a patient has 2 organ system failures, the mortality increases up to 50%. When 3 and more organ systems are malfunctioning, the mortality rises towards 100% [1]. The combination of severe acute respiratory and renal failure is particularly lethal [11]. Although azotemia, fluid and electrolyte balances were well controlled in our patients, the mortality rate was 45%. No death was related to azotemia or its complications, but to progressing or nonresolving MOSF. Mortality increased to 12.5, 16.6 and 57%, respectively, in patients with 2, 3 and >4 organ system failures. Mortality was 100% in patients with MOSF and primary malignancy. There was a significant difference in severity of illness assessed by APSC between survivors and nonsurvivors before starting continuous renal support. Whereas APSC points of survivors fell markedly during the first 4 days of continuous renal support, the APSC of nonsurvivors remained above 20 points. TISS was not significantly different on the first day of therapy.

In conclusion, continuous hemofiltration is an effective method of renal support in critically ill infants and children. It allows good control of azotemia, fluid, electrolyte and acid-base balances. In addition, it allows adequate nutritional support and unlimited medication. Despite effective continuous renal support, the mortality still remains high in children with MOSF. An initial APSC >25 points or a persisting score >20 points combined with intensive therapeutic support (TISS >40 points) are bad prognostic signs for infants and children with ARF.

References

1 Cameron JS: Acute renal failure in the intensive care unit today. Intensive Care Med 1986;12:64–70.
2 Lauer A, Saccaggi A, Ronco C, Belledonne M, Glabman S, Bosch JB: Continuous arteriovenous hemofiltration in the critically ill patient Ann Intern Med 1983;99:455–460.
3 Schetz M, Lauwers PM, Ferdinande P: Extracorporeal treatment of acute renal failure in the intensive care unit: A critical view. Intensive Care Med 1989;15:349–357.
4 Dickson DM, Brown EA, Kox W: Continuous arteriovenous haemodialysis: A new method of complete renal replacement therapy in the critically ill patient. Intensive Care World 1988;5:78–80.
5 Wendon J, Smithies M, Sheppard M, Bullen K, Tinker J, Bihari D: Continuous high-volume venovenous haemofiltration in acute renal failure. Intensive Care Med 1989;15:358–363.
6 Zobel G, Kuttnig M, Ring E, Grubbauer HM: Clinical scoring system in children with continuous renal support. Child Nephrol Urol 1990;10:14–17.
7 Pollack MM, Ruttimann UE, Getson PR: Pediatric Risk of Mortality Score. Crit Care Med 1988;16:1110–1116.
8 Wilkinson JD, Pollack MM, Ruttimann UE: Outcome of pediatric patients with multiple organ system failure. Crit Care Med 1986;14:271–274.
9 Cullen DJ, Civetta JM, Briggs BA: Therapeutic intervention scoring system: A method for quantitative comparison of patient care. Crit Care Med 1974;2:57–60.
10 Hodson EM, Kjellstrand CM, Mauer SM: Acute renal failure in infants and children: Outcome of 53 patients requiring hemodialysis treatment. J Pediatr 1978;93:756–761.
11 Sweet SJ, Glenney CU, FitzGibbon JP, Friedmann P, Teres D: Synergistic effect of acute renal failure and respiratory failure in a surgical intensive care unit. Am J Surg 1981;141:492–496.

Gerfried Zobel, MD, Department of Pediatrics, University of Graz, Auenbruggerplatz, A–8036 Graz (Austria)

Sieberth HG, Stummvoll HK, Kierdorf H (eds): Continuous Extracorporeal Treatment in Multiple Organ Dysfunction Syndrome. Contrib Nephrol. Basel, Karger, 1995, vol 116, pp 169–172

Early Experience with a New Hemodiafilter for Small Infants

G. Zobel, E. Ring, S. Rödl

Department of Pediatrics, University of Graz, Austria

Continuous hemofiltration either driven in the arteriovenous or venovenous mode is a safe and effective extracorporeal treatment modality to control azotemia, fluid and electrolyte balances in critically ill adult as well as pediatric patients with acute renal failure and hemodynamic instability who otherwise cannot tolerate conventional methods of dialytic treatment [1–5]. Although a variety of hemofilters are available for adult and pediatric patients, there are only few hemofilters for neonates and small infants [6–8]. The aim of the study is to show the operational characteristics of a new small hemo(dia)filter for neonates and small infants.

Material and Methods

From June 1993 to June 1994, 8 infants (6 males, 2 females) with a mean age of 7.0 ± 12.3 months and a body weight of 5.7 ± 4.2 kg were treated with continuous extracorporal renal support at the pediatric ICU of the Department of Pediatrics, University of Graz. All infants were on mechanical ventilation and all but 1 needed vasopressor support. Indications for continuous renal support were: acute renal failure combined with multiple organ system failure (n = 6), severe metabolic crisis with hyperammonemia (n = 1), and acute pulmonary edema unresponsive to conventional therapy (n = 1).

The Miniflow 10 (Hospal, Lyon, France) is a new hemo(dia)filter for neonates and small children containing 264 AN 69-HF fibers. The effective length of the fibers is 21 cm, their internal diameter is 240 μm, the membrane thickness is 50 μm, the membrane surface area is 420 cm^2, and the filling volume is 3.7 ml. At the filtrate side there are two ports, one for dialysate inlet and the other one for ultrafiltrate/diafiltrate outlet. In all but 1 patient, hemofiltration was driven in the venovenous mode using a roller pump (Gambro AK 10® blood monitor; Gambro Corp., Hechingen, FRG) with pressure alarms and an air trap,

Table 1. Operational data during continuous hemo-(dia)filtration in 8 infants

Blood flow rate, ml/min	20±2.6
Ultrafiltration rate, ml/min	1.7±0.6
Filtration fraction, %	13±2.4
Mean arterial pressure, mm Hg	58±9.7
Hematocrit, %	35±2.8
Total protein, g/dl	5.3±0.6
Heparin, IU/kg/h	10±4.5
Thrombin time, s	47.8±18.3
Thrombocytes, $10^3/mm^3$	55.7±19.8

Table 2. Operational data during hemofiltration (HF) and hemodiafiltration (HDF) in 3 infants

	HF (n = 23)	HDF (n = 23)	p value
Blood flow, ml/min	20	20	
Ultrafiltration rate, ml/min	1.03±0.1	0.71±0.13	0.06
Urea clearance, ml/min	1.03±0.1	4.3±0.17	0.0001
Creatinine clearance, ml/min	1.09±0.1	4.3±0.18	0.0001
Serum urea, mg/dl	179±9.7	141±5.9	0.002
Serum creatinine, mg/dl	1.4±0.7	1.1±0.1	0.02

n = Number of measurements.

small blood lines and a double-lumen or two single-lumen catheters. In 1 patient the hemofilter was incorporated into the ECMO circuit. The filter was primed with 1 liter of 0.9% NaCl solution containing 5,000 IU of heparin. After priming the blood circuit, the ultrafiltrate or dialysate compartment was primed separately. The ultrafiltrate substitution fluid and the dialysate solution were based on bicarbonate and mixed at the bedside. The dialysate solution was administered in a countercurrent fashion to blood flow at a rate of 3–5 ml/min using the Amicon Equaline® system (Amicon Corp., Lexington, Mass., USA). The dialysate and ultrafiltrate flow were continuously displayed on the control panel of the Equaline® system. Serum and ultrafiltrate/dialysate concentrations of urea and creatinine were determined routinely 3 times/day. Urea clearance was calculated as: Dialysate effluent urea/Serum urea × Volume/Time.

Data are given as mean ± SD. Student's t test for paired and unpaired samples was used for comparison of the mean values.

Informed consent was obtained from the parents of the patients under a protocol approved by the local Ethical Committee.

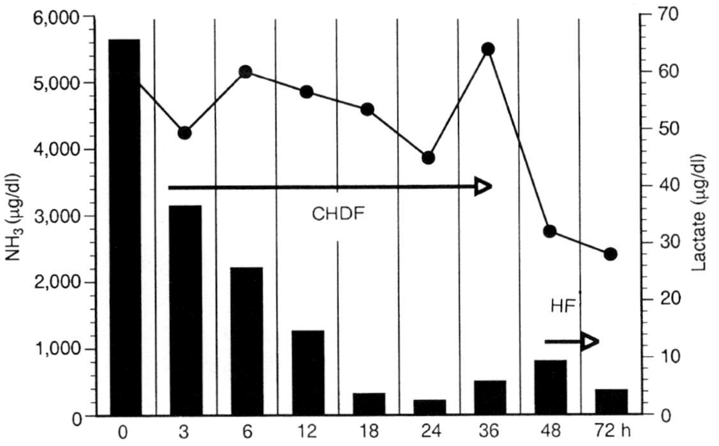

Fig. 1. Serum ammonium (NH_3; ■) and lactate (●) levels during continuous hemodiafiltration (CHDF) in a neonate with severe metabolic crisis due to an inborn error of amino acid metabolism. HF = Hemofiltration.

Results

The mean duration of extracorporeal renal support was 145 ± 187 h, for a total of 1,162 h. The mean hemofilter running time was 53.9 ± 36 h, ranging from 16 to 136 h. The mean prehemofiltration serum urea and creatinine levels were 133 ± 92 and 1.4 ± 0.4 mg/dl, the mean posthemofiltration levels were 100 ± 49 and 1.1 ± 0.4 mg/dl, respectively. The operational data of the new hemo-(dia)filter are given in table 1. Ultrafiltration rates, urea and creatinine clearances during hemofiltration and hemodiafiltration are presented in table 2. Figure 1 illustrates the serum ammonium and lactate levels during hemodiafiltration in a neonate with severe metabolic crisis due to an inborn disorder of amino acid metabolism. In 6 of the 8 patients, correction of fluid and metabolic imbalances was effective and the clinical tolerance to the therapy was excellent. Two of these patients died later in the course of their illness and 2 patients with severe low cardiac output syndrome and multiple organ system failure died within 6 h of extracorporeal treatment.

Discussion

Continuous hemofiltration or hemodiafiltration either driven in the arteriovenous or venovenous mode has become a good alternative to peritoneal dialysis or may even represent a first-choice treatment in critically ill infants with acute renal failure or multiple organ system failure. The hemofilter device for small infants must have a low priming volume, should operate at low flow and pressures, and should be tolerated by critically ill infants with hemodynamic instability.

The polyacrylonitrile AN 69 membrane showed excellent filter performance in critically ill adult patients [9–11]. In our infants the new hemo(dia)filter worked excellently in the pump-driven mode in terms of ultrafiltration rates and filter running time. The AN 69-HF fibers and the special geometry of the filter allowed a significant increase of the urea and creatinine clearances during hemodiafiltration. In addition, the clinical tolerance to the hemo(dia)filter was excellent.

References

1 Kramer P, Wigger W, Rieger J, Mathaei D, Scheler F: Arteriovenous hemofiltration: A new and simple method for the treatment of overhydrated patients resistant to diuretics. Klin Wochenschr 1977;55:1121–1122.

2 Wendon J, Smithies M, Sheppard M, Bullen K, Tinker J, Bihari D: Continuous high volume venovenous haemofiltration in acute renal failure. Intensive Care Med 1989;15:358–363.

3 Lauer A, Saccaggi A, Ronco C, Belledonne M, Glabman S, Bosch JB: Continuous arteriovenous hemofiltration in the critically in patient. Ann Intern Med 1983;99:455–460.

4 Lieberman KV: Continuous arteriovenous hemofiltration in children. Pediatr Nephrol 1987;1: 330–338.

5 Leone MR, Jenkins RD, Golper TA, Alexander SR: Early experience with continuous arteriovenous hemofiltration in critically ill pediatric patients. Crit Care Med 1986;14:1058–1063.

6 Ronco C, Brendolan A, Bragantini L, Chiaramonte S, Feriani M, Fabris A, DellAquila R, La Greca G: Treatment of acute renal failure in newborns by continuous arteriovenous hemofiltration. Kidney Int 1986;29:908–915.

7 Zobel G, Kuttnig M, Ring E: Continuous arteriovenous hemodialysis in critically ill infants. Child Nephrol Urol 1990;10:196–198.

8 Zobel G, Ring E, Müller WD: Continuous arteriovenous hemofiltration in preterm infants. Crit Care Med 1989;17:534–536.

9 Dickson DM, Brown EA, Kox W: Continuous arteriovenous haemodialysis: A new method of complete renal replacement therapy in the critically ill patients. Invenstive Care World 1988;5:78–80.

10 Reynolds HN, Borg U, Belzberg H, Wiles C: Efficacy of continuous arteriovenous hemofiltration with dialysis in patients with renal failure. Crit Care Med 1991;19:1387–1394.

11 Belomo R, Tipping P, Boyce N: Continuous veno-venous hemofiltration with dialysis removes cytokines from the circulation of septic patients. Crit Care Med 1993;21:522–526.

Gerfried Zobel, MD, Department of Pediatrics, University of Graz, Auenbruggerplatz, A-8036 Graz (Austria)

Sieberth HG, Stummvoll HK, Kierdorf H (eds): Continuous Extracorporeal Treatment in Multiple Organ Dysfunction Syndrome. Contrib Nephrol. Basel, Karger, 1995, vol 116, pp 173–178

Continuous Venovenous Hemofiltration in a Seven-Year-Old Child Suffering from Severe Sepsis and Multiple Organ Failure

R. Lampert[a], *E.H. Weih*[a], *R. Hikl*[a], *M. Mazuch*[b]

[a] Institut für Anästhesie und Intensivmedizin und
[b] Klinik für Allgemeinchirurgie, Städtische Kliniken Wuppertal/
Universität Witten/Herdecke, Wuppertal, Germany

Recently, the efforts to eliminate endotoxins and mediator substances in septic patients by employment of continuous venovenous hemofiltration (CVVH) have often been discussed and the improvement of the patients' general clinical condition as a result of this therapy has been described [1, 2]. In the following case report, the use of CVVH in a 7-year-old child suffering from severe sepsis and coagulation disturbance is described. In our opinion the patient profited from the use of this method in respect of hemostasis.

Case Report

Four weeks after an appendectomy because of a perforated appendicitis and a Douglas' abscess, a 7-year-old child, 28 kg in weight, received an emergency laparotomy to treat a protracted mechanical ileus. As a part of the operation, adhesions were released and a manual decompression of the ileum was carried out. Striking was the postoperative 2-hourly passage of protein-rich feces with a loss of up to 2,500 ml of intestinal content per day. On the second postoperative day the recurrence of ileus symptoms necessitated a relaparotomy. This disclosed an enormous obstructed, edematous ileum, which was again manually decompressed and relieved by application of a double-bore preternatural sigmoidal anus. Postoperatively the child developed a severe sepsis (temperature >39 °C, leukocyte count >20,000/μl, thrombocytes drop from 340,000 to 45,000/μl, lactate 3.8 mmol/l). Blood cultures revealed the repeated presence of *Enterococcus; Proteus vulgaris* was detected in tracheal secretion and gastric juice. These bacteria were treated with specific antibiotics. Over the next 4 days a capillary leakage syndrome caused an increase in weight of 6 kg. Unfortunately an analysis of the level of endotoxins in the blood was not carried out. Over the following days the serous fluid loss out of the preternatural anus and the abdominal drain reached up to 2,800 ml/day with a maximum protein content of 5.0 g/dl. The thrombocyte count fell to 17,000/μl. Petechial bleeding appeared on the body trunk and there was profuse bleeding from the mucous membranes of mouth and nose. Protein coagulation system was equally extremely pathological under substitution (PTT >3 min, Quick 49%, AT3 29%, factor VIII 15%, factor IX 25%, fibrinogen <1.0 g/l, fibrinogen fission products 10–40 μg/ml). Within 10 days we transfused 46 units of thrombocyte concentrates, 29 units of fresh-frozen plasma and replaced clotting factors (8,000 IU AT3, 1,000 IU PPSB, 1,250 IU factor VIII) to stabilize the coagulation. An enormous drop in immunoglobulins went along with the loss of clotting factors. To raise the IgG fraction from a level of 155 mg% into the normal range (770–1,510 mg%) the substitution of a total of 105 g immunoglobulins over 8 days was necessary.

An unchanged septic picture and the suspicion of intra-abdominal bleeding necessitated a further operative review on the 11th postoperative day. The ileum was seen to be much less edematous. Neither an intra-abdominal abscess nor a source of bleeding were discovered. After the operation an attempt was made, using a pump-operated venovenous hemofiltration (support apparatus: Blood Monitor BM11, Dialysetechnik; Hemofilter: AV 400; Fresenius), to hinder the progression of the illness while retaining diuresis and normal retention levels. With an average pump speed of 80 ml/min, 4.8 liters of blood, more than double the child's blood volume, were run through the blood filter. The use of the CVVH made it possible to extract, on average, 800 ml ultrafiltrate/h. With this procedure there was an average daily fluid turnover of about 20 liters with a negative balance of 1–2 liters obtained.

First visible effects were the fall of the body temperature to a stable 37 °C and an improvement in coagulation within 24 h. The PTT fell from an average level of 140 to 100 s (fig. 1). Because of coagulation in the hemofilters, careful heparinization was necessary. The application was regulated according to PTT levels and had to be interrupted intermittently. The thrombocyte level decreased again to 42,000/μl over the following days, but platelet substitution could now be avoided (fig. 2). The protein loss and thereby the necessity of transfusion of fresh-frozen plasma and albumin diminished considerably.

The attempts to end CVVH early resulted in reproducible fulminant regressions of the general condition of the child. On the 10th day of filtration the complication of a 3-day-long oligoanuria developed, which eventually resolved spontaneously. A total of 17 days of hemofiltration was necessary. During this time we had to change the filter 19 times, 14 times due to a thrombosis in the hemofiltration system, 5 times due to a grave reduction of ultrafiltrate levels. The child's general condition improved gradually. A further operative revision was not necessary. From the 25th postoperative day the child passed feces regularly over the preternatural anus, from the 27th day we started with enteric feeding. On the 32nd day the withdrawal of mechanical ventilation was possible. On the 42nd day we were able to transfer the child to a regular surgical ward.

Discussion

This case describes a severely septic condition as the result of a long-standing ileus with an accompanying disturbance in coagulation, pulmonary and renal function. Microcirculatory disturbances of the bowel segment in question lead to

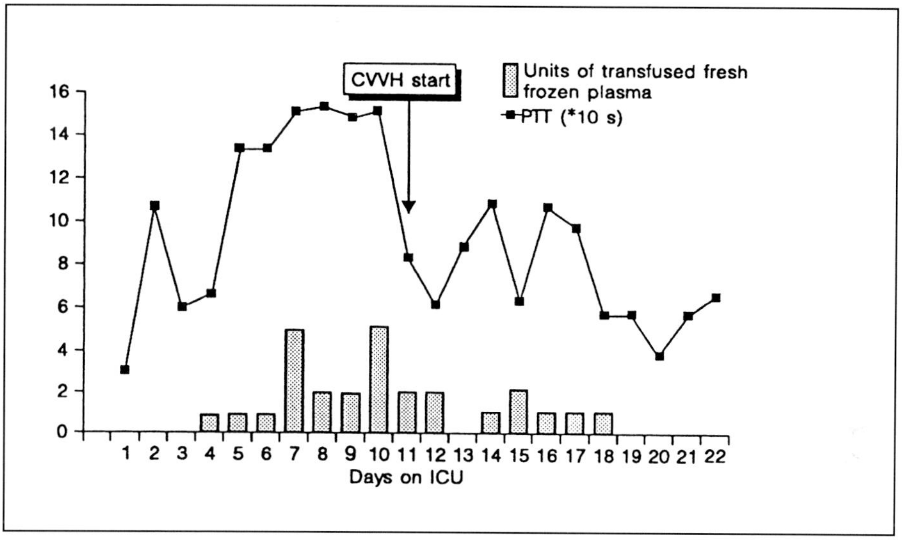

Fig. 1. The drop of the PTT level and the distinct decrease of transfused FFP after the beginning of hemofiltration. The strong decrease of the PTT on the 10th day may be due to a peritoneal dialysis which had to be stopped after 1 day because of technical problems.

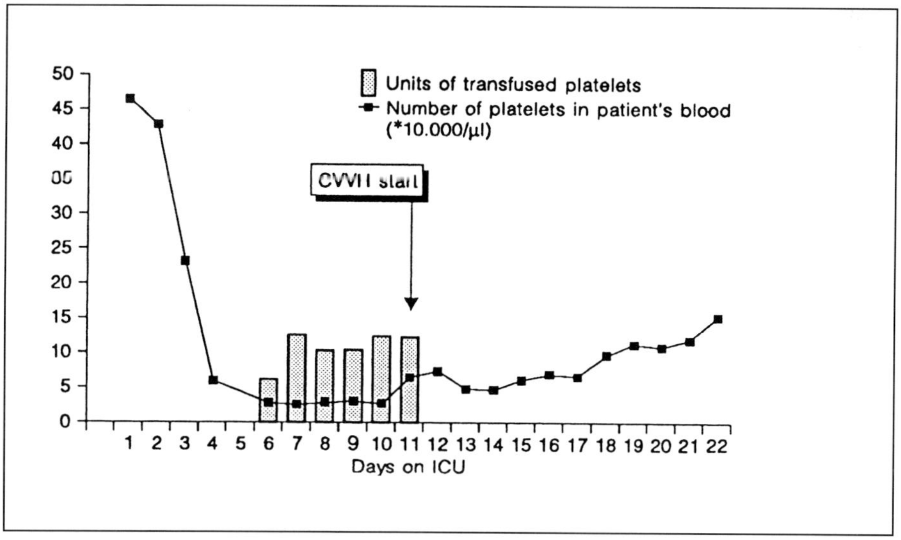

Fig. 2. Stabilization of the thrombocyte count without further substitution after starting CVVH.

hypoxic endothelial swelling and result in damage to the entire intestinal wall. The decompression of the bowel goes along with the development of a reperfusion syndrome [3, 4]. The resulting amount of free oxygen radicals leads to an increasing destruction of cell and membrane proteins and encourages the aggregation and adherence of polymorphonuclear leukocytes to the vascular endothelium. The increasing damage of the endothelium leads to an increased flow of fluid into the cells and an ensuing edema of the intestinal wall, which eventually allows a large translocation of microorganisms and endotoxins from the intestinal lumen to portal circulation. Endotoxins cause the release of cytokines, for example interleukins, tumor necrosis factor (TNF) and products of C20-arachidonic acid. One effect of these substances is on the hemostasis system. TNF and endothelium-activating factor (EAF) alter the structure of the endothelial lining and lead to adhesions of thrombocytes and to an activation of the clotting process. Thromboxane A_2 and platelet-activating factor (PAF) favor aggregation of thrombocytes and leukocytes [5]. These cell aggregates activate the intrinsic clotting system. TNF and interleukin-1 cause an increase of plasminogen activator inhibitor (PAI) and therefore an inhibition of fibrinolysis. The total effect of the mediators of sepsis on the clotting system is a state of hypercoagulation [6, 7]. There are some clinical observations which give reason to hope that the employment of CVVH can favorably influence the course of severe sepsis [1, 2, 8]. The straight forward possibility of fluid withdrawal results in an avoidance of generalized edema. The decrease in extracellular fluid volume leads to a smaller diffusion route between the capillary bed and the cells. The oxygen transport to the tissues is made easier and regeneration processes are furthered. This could explain the improvement in organ function but also possibly the avoidance of translocation due to the decreased edema of the intestinal wall. A further conceivable influence of hemofiltration is the elimination of cytotoxic substances with low molecular weights, for example derivatives of arachidonic acid or kinins. High molecular endotoxin, however, cannot pass the filter (table 1). A further mechanism that facilitates the removal of sepsis mediators from the bloodstream is the absorption of these substances by the filtration membrane [9].

In our case there was no necessity of transfusion of platelets and clotting factors after starting CVVH and we were able to reduce the substitution of fresh-frozen plasma and albumins. A plausible explanation could be, on the one hand, the reduction of edema in the intestinal wall and a reduced translocation due to the fluid withdrawal in the filtration process and, on the other hand, through the elimination of clotting-activating mediators. By removing thromboxane A_2 and PAF the readiness of the platelets to aggregate is reduced and a stablization of the thrombocyte count is possible. This is an astonishing phenomenon under CVVH because normally there is evidence of platelet loss on the foreign synthetic surface membrane. A side effect of the extracorporeal circulation is the reduction in body

Table 1. Molecular weight of some sepsis mediators

Mediators of sepsis	Molecular weight, daltons
Thromboxane A_2	352
Prostaglandins	≈350
Leukotrienes	≈600
Platelet-activating factor	524
Lipid A	≈2,000
Interleukin-1, -2	≈15,000
Interleukin-6	≈25,000
Tumor necrosis factor	≈17,000
Albumin	≈69,000
Endotoxin	≈100,000

It is of clinical relevance that particles up to 5,000 dalton molecular weight easily pass the filter, whereas for a significant elimination of larger particles an increase of the filtration rate is necessary. For very large particles such as endotoxin, the adsorption at the filter surface is discussed.

temperature by approximately 1–1.5 °C. Thus the oxygen demand of the organism is reduced and the disturbed balance in oxygen supply and demand is improved.

In view of the blood loss resulting from 19 changes of the hemofiltration system, a discontinuation of this therapy has been discussed daily. Even though measurable parameters could not actually prove the favorable effects of CVVH, the stabilization of the child's clinical condition was clearly evident to the doctors involved. Each interruption in CVVH produced a deterioration in the general condition of the patient which immediately resolved with restarting hemofiltration.

Conclusion

CVVH first of all is a symptom-orientated therapy which is able to decelerate the septic process. Specific complications of hemofiltration are seldom to be observed, when implemented by experienced operators, so that CVVH can be seen as a bearable additional risk. Nevertheless, this technique should only be employed in cases where operative procedures and intensive care produce no further improvement, while the patient is still vitally endangered. In how far this technique has influence on the lethality of a severe sepsis can only be proved in clinical studies on a larger scale.

References

1 Bellomo R, Tipping P, Boyce N: Continuous veno-venous hemofiltration with dialysis removes cytokines from circulation of septic patients. Crit Care Med 1993;21:522–526.

2 Ossenkoppele GJ, van der Meulen J, Bronsveld W, Thijs LG: Continuous arterio-venous hemofiltration as an adjunctive therapy for septic shock. Crit Care Med 1985;13:102–104.

3 Hamelmann H, Deltz E, Beske C: Der Dickdarmileus. Chirurg 1989;60:185–188.

4 Zühlke HV, Lorenz EPM, Harnoss BM, Häring R, Rodloff A: Endotoxinämie und Bakteriämie unter manueller oraler Dekompression im Ileus. Chirurg 1988;59:349–356.

5 Seeger W, Lasch HG: Septic lung. Rev Infect Dis 1987;9:570–579.

6 Biemond BJ, Levi M, van Deventer SJH: The procoagulant state in sepsis: Experimental models and new modalities for intervention; in Lamy M, Thijs LG (eds): Mediators of Sepsis. Berlin, Springer, 1992, pp 113–123.

7 Van der Poll T, Levi M, van Deventer JH: Tumor necrosis factor and the disbalance between coagulant and anticoagulant mechanisms in septicemia; in Vincent LJ (ed): Update 1991. Update in Intensive Care and Emergency Medicine. Berlin, Springer, 1991, vol 14, pp 269–273.

8 Barckow D: Mediatorenelimination (Hämofiltration, Plasmapherese, Hämoperfusion); in Schuster HP (ed): Intensivtherapie bei Sepsis und Multiorganversagen. Berlin, Springer, 1993, pp 132–149.

9 Van der Mer JWM, Blok WL, Barrera P: Ways to interfere with lethal cytokinemia; in Vincent LJ (ed): Update 1991. Update in Intensive Care and Emergency Medicine. Berlin, Springer, 1991, vol 14, pp 245–253.

Dr. med. Reinhard Lampert, Institut für Anästhesie und Intensivmedizin,
Kliniken der Stadt Wuppertal, Heusnerstrasse 40, D–42283 Wuppertal (Germany)

Sieberth HG, Stummvoll HK, Kierdorf H (eds): Continuous Extracorporeal Treatment in Multiple Organ Dysfunction Syndrome. Contrib Nephrol. Basel, Karger, 1995, vol 116, pp 179–182

Removal of Complement Fragments by Ultrafiltration during Pediatric Cardiopulmonary Bypass

Didier Journois[a], *Laetitia Vaccaroni*[a], *Régis Rambur*[b], *Michel Lagarde*[b], *Moise Keita*[b], *Philippe Pouard*[a]

Departments of
[a] Anesthesia and Intensive Care Medicine, and
[b] Pediatric Cardiovascular Surgery, Laennec Hospital, Paris, France

Andreasson et al. [1] have recently confirmed the activation of the complement cascade by cardiopulmonary bypass (CPB) during pediatric cardiac surgery and that maximal complement release coincides with the period of rewarming. Furthermore, C3a and C5a have been shown to be responsible for the delayed release of cytokines following CPB. The current prospective study was intended to determine the effects of ultrafiltration on complement fraction removal.

Material and Methods

Thirty-two children undergoing surgical correction of tetralogy of Fallot were randomly assigned to be ultrafiltered during CPB rewarming or a control group. Ultrafiltration was performed with a polysulfone hemofilter (HFT02®; Sorin Lab.) with a rate adjusted to reach a cardiotomy reservoir level of zero at the completion of rewarming. Complement fractions C3a and C5a were radioimmunoassayed before and after rewarming ± ultrafiltration (T_2 and T_3; Amersham, France). The ratio of ultrafiltrated volume by the estimated total blood volume (UF/TBV) was calculated to compare the relative amount of ultrafiltrate withdrawn in patients. TBV was defined as prime volume plus patient blood volume [3]. This ratio allowed to study the correlation between UV and C3a and C5a changes. Data were analyzed using the Mann-Whitney test for unpaired data. Correlations were studied with the nonparametric Spearman rank correlation test. Statistical significance was chosen at $p < 0.05$.

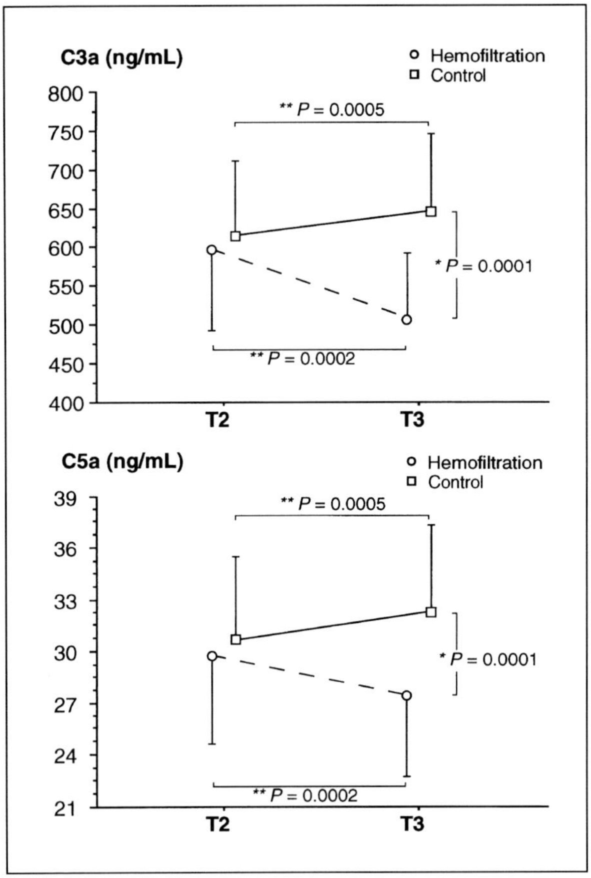

Fig. 1. Evolution of the complement fragments C3a and C5a during the period of rewarming ± hemofiltration in the two groups.

Results

The plasma concentrations of C3a and C5a before rewarming were 605 ± 98 and 30 ± 4.5 ng/ml respectively for both groups. Plasma levels of these fractions were significantly different between T_2 and T_3 ($p < 0.0001$) and differed between groups (fig. 1). For both C3a and C5a, a significant increase occurred during rewarming in the control group while a decrease was observed in the ultrafiltration group. The complement fraction variations correlated with the UF/TBV ratio (fig. 2).

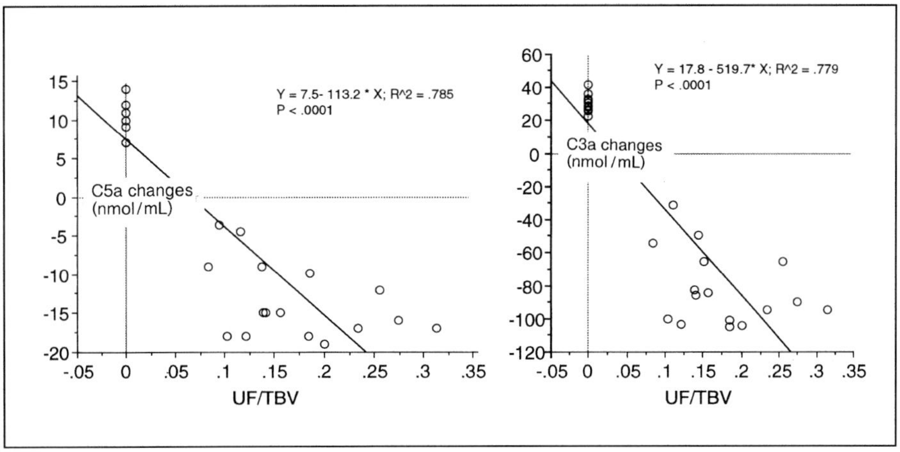

Fig. 2. Correlation between changes in complement fragment plasma concentrations during hemofiltration ± rewarming and the amount of ultrafiltrate withdrawn in each patient expressed as the UF/TBV ratio.

Discussion

The present study showed that ultrafiltration actually reduced the plasma concentrations of complement components. Moreover, this removal seemed to be linearly correlated with the amount of withdrawn ultrafiltrate. This phenomenon confirms the existence of a convective removal of complement fractions by ultrafiltration that was already suggested by the presence of these components in the ultrafiltrate [1]. Despite their removal, the observed levels of C3a and C5a still remain higher than normal. Additional complement activation produced by the ultrafiltration circuit may occur during the use of ultrafiltration membranes [4]. This could lead to prefer polyacrylonitrile hemofilters which are known to induce a lesser complement activation than the polysulfone hemofilters used in this study [4]. An important aspect of C3a and C5a removal might be the limitation of the delayed effects that these components have on cytokines release [2]. These results suggest that ultrafiltration may be useful during CPB rewarming to limit the effects of CPB on the inflammatory cascade.

References

1 Andreasson S, Göthberg S, Berggren H, Bengtsson A, Eriksson E, Risberg B: Hemofiltration modifies complement activation after extracorporeal circulation in infants. Ann Thorac Surg 1993;56: 1515–1517.

2 Haeffner-Cavaillon N, Roussellier N, Ponzio O, Carreno MP, Laude M, Carpentier A, Kazatchkine MD: Induction of interleukin-1 production in patients undergoing cardiopulmonary bypass. J Thorac Cardiovasc Surg 1989;98:1100–1106.

3 Naik SK, Knight A, Elliott M: A prospective randomized study of a modified technique of ultrafiltration during pediatric open-heart surgery. Circulation 1991;84(suppl III):422–431.

4 Mulvihill J, Cazenave JP, Mazzucotelli JP, Crost T, Collier C, Renaux JL, Pusineri C: Minimodule dialyser for quantitative ex vivo evaluation of membrane haemocompatibility in humans: Comparison of acrylonitrile copolymer, cuprophan and polysulphone hollow fibres. Biomaterials 1992;13: 527–536.

Dr. D. Journois, Hôpital Laennec, 42, rue de Sèvres, F–75007 Paris (France)

Subject Index

Acetaminophen, renal toxicity 22
Acetate, vasodilation 25, 45
Activated clotting time
continuous venovenous hemofiltration filter running time, effects 145, 147, 148
measurement 30
Acute renal failure
continuous hemofiltration 125, 126
incidence in open heart surgery 136
mortality with multiple organ dysfunction syndrome 167, 168
prognostic factors 34, 36
renal replacement therapy goals 38, 62, 163
Adult respiratory distress syndrome
continuous arteriovenous hemofiltration 126
endothelial cell necrosis 1
pulmonary function and continuous venovenous hemofiltration 108–111
systemic inflammatory response syndrome pathogenesis 80
AN 69, *see* Polyacrylonitrile membrane
Anticoagulation, *see also* Heparin
blood-air contact effects 134
coagulation marker assay 140–142
thromboelastography monitoring 159–162
Antithrombin III, *see also* Thrombin-antithrombin III complex
deficiency during continuous venovenous hemodialysis 154–157
hepatic synthesis 156
Apache II score
dialysis mode effects 34–36
prognostic value 36, 100
Azotemia, anion buffer effects 42

Bicarbonate
acidosis control 43, 46
comparison to lactate 38–46
stability of solutions 38, 44

C3a, *see* Complement compounds
C5a, *see* Complement compounds
Cardiac output, monitoring during hemodialysis 112–114
Cardiopulmonary bypass
capillary leakage 106
complement fragment removal 82, 179–181
cytokine
release 81, 82, 87, 88
removal 82–84, 86–88
systemic inflammatory response syndrome pathogenesis 80
ultrafiltration effects on hemostasis 150–153
Cell adhesion molecule
induction by endotoxins 7
regulation by endothelium 4

Cerebral edema, intensive care management 23
Complement compounds
C3 convertase 80, 81
hypothermia and reduced activation 81
immunoassay 179
neutropenia role 81
reactive oxygen species induction 81
removal
continuous venovenous hemofiltration 76–79, 82, 116
ultrafiltration 179–181
Continuous arteriovenous hemodiafiltration
cytokine removal 62, 63, 65, 67, 69–74
systemic inflammatory response syndrome treatment 69–74
ultrafiltrate clearance 70
Continuous arteriovenous hemofiltration
children 163, 164
cytokine removal 124, 125
nonrenal indications for treatment 121, 122, 127
sepsis therapy 121–124
Continuous venovenous hemodiafiltration
Apache III score, effect in sepsis 91–93, 95
body temperature effects 13
cytokine removal 11–13, 89–96
methotrexate removal from blood 48, 49, 51, 52
thromboelastography monitoring of coagulation 159–162
Continuous venovenous hemodialysis
antithrombin III deficiency 154–157
intermittent hemodialysis comparison 34–36
Continuous venovenous hemofiltration
anion buffer selection 38, 40–46
cardiac stability in sepsis 99–101, 103–106
children 163, 164, 173–178
complement compound removal 76–79, 82, 116
cytokine removal 56, 58, 59, 76–79, 99–101
filter running time and coagulation parameters 145, 147, 148
hemostasis effects, comparison to intermittent hemodialysis 140–144
lipid peroxidation in patients 58–60
multiple organ dysfunction syndrome with normal renal function, treatment benefits 115, 116, 118, 119
necrotizing pancreatitis therapy 115, 118, 119
nephrotic syndrome treatment 53–55
nonrenal indications for treatment 121, 122, 127
oxygen extraction rate 57, 58
platelet effects 176
pulmonary function, effect in sepsis 108–111
systemic inflammatory response syndrome response 57–60
thrombocytopenia treatment 136–139
thromboelastography monitoring of coagulation 159–162
Convection, solute transport 12, 13
Cuprophane membrane
neutrophil activation 13
permeability coefficient 29

Diffusion, solute transport 12

Electrolyte, anion buffer effect on serum level 41, 42, 44
Endothelial cell
endotoxin damage 6, 176
growth factor synthesis 6
inflammation amplification 4, 5
microcirculatory dysfunction in multiple organ dysfunction syndrome 2, 7, 8
thrombogenic activity 2, 3
vascular tone regulation 3, 4
Endotoxin
assay 90, 109, 110
cytokine release stimulation 176
membrane adsorption 126
Epoprostenol, anticoagulation and dialysis 24, 25
Extracorporeal circuit, modes 32

Fibrinopeptide A, coagulation marker assay 140–142

Fulminant hepatic failure
circulatory disturbances 22
intensive care management
anionic base, dialysate 25
cerebral edema 23
extracorporeal anticoagulation 24, 25
renal replacement therapy 23, 24
renal failure association 22

Hemodiafiltration, *see also* Continuous arteriovenous hemodiafiltration, Continuous venovenous hemodiafiltration
definition 31
infant filter 169–172
modes 31
Hemodialysis, *see also* Continuous venovenous hemodialysis, Intermittent hemodialysis
cardiac output monitoring 112–114
definition 29, 31
indications 32
low-dose heparinization 130, 131, 133–135
modes 31
Hemofiltration, *see also* Continuous arteriovenous hemofiltration, Continuous venovenous hemofiltration
complications of continuous treatment 126, 127
definition 31
infant filter 169, 172
low-dose heparinization 130, 131, 133–135
modes 31
Heparin
anticoagulation
dialysis 24
low molecular weight form, advantages 159
ultrafiltration 151
antithrombin III deficiency role 156, 157
continuous hemofiltration and low-dose heparinization 130, 131, 133–135
filter running time and coagulation parameters, continuous venovenous hemofiltration 145, 147, 148
membrane adsorption 133, 134
platelet factor 4 neutralization 148
thrombocytopenia association 136
Hepatic failure, *see* Fulminant hepatic failure
High-flux dialysis, definition 31

Interleukin-1
immunoassay 65, 86
removal
continuous hemodiafiltration 63–65, 67, 69–74
continuous ultrafiltration 82, 83, 86–88
filter duration 70, 71
mechanism 74
Interleukin-6
immunoassay 65, 86, 90, 101
removal
continuous hemodiafiltration 63–65, 67, 69–74, 95, 96
continuous venovenous hemofiltration 76–79, 100, 101
filter duration 70–72
mechanism 74
ultrafiltration 86–88
Interleukin-8
immunoassay 86
levels in systemic inflammatory response syndrome 11, 88
removal by continuous ultrafiltration 83, 86–88
sepsis role 82
Intermittent hemodialysis
continuous venovenous hemodialysis comparison 34–36
hemostasis effects, comparison to continuous venovenous hemofiltration 140–144
Intracranial pressure, intensive care management 23, 25, 26

Lactate
acidosis control 43, 46
cardiovascular effects 45
comparison to bicarbonate 38–46
hepatic clearance rate 25

Lactate (continued)
patient load in dialysis 38, 45, 46
protein catabolism effects 45, 46
stability of solutions 38
Lipid peroxidation, continuous venovenous hemofiltration effects 58–60

Malondialdehyde, lipid peroxidation marker 58–60
Mannitol, intracranial pressure management 25, 26
Methotrexate
cytotoxicity 48
removal from blood 48, 49, 51, 52
tissue storage 51
Miniflow 10 filter
clearance rates 171
infant filter 169–172
size 169
Multiple organ dysfunction syndrome
antisense oligonucleotide therapy 7
continuous venovenous hemofiltration
children 163, 164, 173–178
treatment benefits with normal renal function 115, 116, 118, 119, 127, 128
endothelial dysfunction 2, 7, 8
hemofiltration in children 163–168
immunotherapy 7
latency period 1
mortality 163, 167, 168
priming of inflammation 4
prognostic factors
children 163–168
sepsis 18–20
scoring systems
Apache II scoring 18, 34, 35
children 163, 167, 168

Necrotizing pancreatitis, continuous venovenous hemofiltration therapy 115, 118, 119
Nephrotic syndrome
complications 53
hemofiltration therapy 53–55
proteinuria
management 53
values 53

Pancreatitis, *see* Necrotizing pancreatitis
Partial thromboplastin time
continuous venovenous hemofiltration
filter running time, effects 145, 147, 148
measurement 30
Platelet
membrane contact and activation 144
transfusion 176
Polyacrylonitrile membrane
biocompatibility 24
cytokine removal 63, 65, 67, 73
permeability coefficient 29
Propofol, intracranial pressure management 26
Prostacyclin, thrombocytopenia treatment 136–139

Reactive oxygen species, induction by complement compounds 81

Sepsis
adhesion molecule expression 5
antibiotic therapy 16
bacteria species and outcome 19
capillary leakage 104, 106
continuous arteriovenous hemofiltration 121–124
continuous hemodiafiltration and sepsis survival 11–13
continuous venovenous hemofiltration
cardiac stability effects 99–101, 103–106, 123, 124
children 163, 164, 173–178
pulmonary function effects 108–111, 123
cytokine
antagonists in treatment 11
mediation 10, 94, 99
endothelial dysfunction 2
endotoxin response of endothelial cells 6, 7
fever and outcome 19
immunotherapy 17, 19, 94
incidence 16, 17, 94
mortality 16, 17, 110
prognosis in intensive care units 17–19

scoring systems 18
vascular tone response 3
vasodilation 99
Serotonin, endotoxin assay 109, 110
Systemic inflammatory response syndrome
association with multiple organ dysfunction syndrome 56
capillary leakage 104, 106, 150
circulatory abnormalities 56, 57
clinical manifestations 63
continuous arteriovenous hemodiafiltration 63–65, 67, 69–74
continuous venovenous hemofiltration 57–60, 127, 128
cytokine levels 10, 11
endothelial dysfunction 2

Terminal complement complex
immunoassay 90
removal from blood 93, 95, 96
Thiopentone, intracranial pressure management 26
Thrombin-antithrombin III complex, coagulation marker assay 140–142
Thrombocytopenia
continuous venovenous hemofiltration treatment 116
heparin association 136
prostacyclin treatment 136–139
Thromboelastography, coagulation monitoring 159–162
β-Thromboglobulin, coagulation marker assay 140–142
Thromboxane, removal by continuous arteriovenous hemofiltration 122
Transmembrane pressure gradient, solute transport 13
Tumor necrosis factor-α
immunoassay 65, 86, 90, 101
levels in systemic inflammatory response syndrome 10, 11
oligomerization 12, 95
removal
continuous hemodiafiltration 12, 32, 63–65, 67, 69–74, 92, 95
continuous hemofiltration 59, 76–79, 100, 101, 125
continuous ultrafiltration 82, 83, 86–88
filter duration 70, 71
mechanism 74, 95

Ultrafiltration
complement compound removal 179–181
cytokine removal 82–88
definition 31
effect on hemostasis in cardiopulmonary bypass 150–153
indications 32
modes 31